VOLUME ONE HUNDRED AND THREE

ADVANCES IN PARASITOLOGY

Highlighting Operational and Implementation Research for Control of Helminthiasis

VOLUME ONE HUNDRED AND THREE

Advances in PARASITOLOGY

Highlighting Operational and Implementation Research for Control of Helminthiasis

Edited by

JENNIFER KEISER
Head, Helminth Drug Development Unit, Department of Medical Parasitology and Infection Biology, Swiss Tropical and Public Health Institute, Basel, Switzerland

ELSEVIER

ACADEMIC PRESS
An imprint of Elsevier

Academic Press is an imprint of Elsevier
125 London Wall, London EC2Y 5AS, United Kingdom
525 B Street, Suite 1650, San Diego, CA 92101, United States
50 Hampshire Street, 5th Floor, Cambridge, MA 02139, United States
The Boulevard, Langford Lane, Kidlington, Oxford OX5 1GB, United Kingdom

First edition 2019

ISBN: 978-0-08-102750-9
ISSN: 0065-308X

For information on all Academic Press publications visit our website at https://www.elsevier.com/books-and-journals

Publisher: Zoe Kruze
Acquisition Editor: Ashlie M. Jackman
Editorial Project Manager: Ana Claudia Garcia
Production Project Manager: Abdulla Sait
Cover Designer: Victoria Pearson

Typeset by TNQ Technologies

CONTENTS

CONTRIBUTORS

Heather Melanie R. Ames
Norwegian Institute of Public Health, Division for Health Services, Oslo, Norway; Swiss Tropical and Public Health Institute, Basel, Switzerland; University of Basel, Basel, Switzerland

Deon Bezuidenhout
Merck (Pty) Ltd, South Africa, an affiliate of Merck KGaA, Darmstadt, Germany

Isaac I. Bogoch
Department of Medicine, University of Toronto, Toronto, ON, Canada; Divisions of General Internal Medicine and Infectious Diseases, Toronto General Hospital, University Health Network, Toronto, ON, Canada

Xavier Bosch-Capblanch
Swiss Tropical and Public Health Institute, Basel, Switzerland; University of Basel, Basel, Switzerland

Jong-Yil Chai
Institute of Parasitic Diseases, Korea Association of Health Promotion, Seoul, Republic of Korea; Department of Parasitology and Tropical Medicine, Seoul National University College of Medicine, Seoul, Republic of Korea

N'da Angbeletchi David Aka
Université Félix Houphouët-Boigny, Abidjan, Côte d'Ivoire

Thomas Fürst
Swiss Tropical and Public Health Institute, Basel, Switzerland; University of Basel, Basel, Switzerland; School of Public Health, Imperial College London, London, United Kingdom

Eric Huber
Swiss Tropical and Public Health Institute, Basel, Switzerland; University of Basel, Basel, Switzerland

Rekol Huy
National Center for Parasitology, Entomology and Malaria Control, Phnom Penh, Cambodia

Jennifer Keiser
Swiss Tropical and Public Health Institute, Basel, Switzerland; University of Basel, Basel, Switzerland

Virak Khieu
National Center for Parasitology, Entomology and Malaria Control, Phnom Penh, Cambodia

Charles H. King
Center for Global Health and Diseases, WHO Collaborating Centre for Research and Training on Schistosomiasis Elimination, Case Western Reserve University School of Medicine, Cleveland, OH, United States; Schistosomiasis Consortium for Operational Research and Evaluation (SCORE), University of Georgia, Athens, GA, United States

Jonathan D. King
Department of Control of Neglected Tropical Diseases, World Health Organisation, Geneva, Switzerland

Elly Kourany-Lefoll
Ares Trading S.A., Switzerland, an affiliate of Merck KGaA, Darmstadt, Germany

Hanspeter Marti
Swiss Tropical and Public Health Institute, Basel, Switzerland; University of Basel, Basel, Switzerland

Kazuko Miyamoto
School of Nursing, Faculty of Medicine & Center for International Education and Exchange, Yamanashi University, Yamanashi, Japan

Wendelin Moser
Swiss Tropical and Public Health Institute, Basel, Switzerland; University of Basel, Basel, Switzerland

Sinuon Muth
National Center for Parasitology, Entomology and Malaria Control, Phnom Penh, Cambodia

Eliézer N'Goran
Université Félix Houphouët-Boigny, Abidjan, Côte d'Ivoire

Peter Odermatt
Swiss Tropical and Public Health Institute, Basel, Switzerland; University of Basel, Basel, Switzerland

Mamadou Ouattara
Université Félix Houphouët-Boigny, Abidjan, Côte d'Ivoire

Christian Schindler
Swiss Tropical and Public Health Institute, Basel, Switzerland; University of Basel, Basel, Switzerland

Peter Steinmann
Swiss Tropical and Public Health Institute, Basel, Switzerland; University of Basel, Basel, Switzerland

J. Russell Stothard
Department of Parasitology, Liverpool School of Tropical Medicine, Liverpool, United Kingdom

Alon Vasiman
Department of Medicine, University of Toronto, Toronto, ON, Canada

Tai-Soon Yong
Department of Environmental Biology, Institute of Tropical Medicine, and Arthropods of Medical Importance Resource Bank, Yonsei University College of Medicine, Seoul, Republic of Korea

Meike Zuske
Swiss Tropical and Public Health Institute, Basel, Switzerland; University of Basel, Basel, Switzerland

PREFACE

On December 7–8, 2017 the Swiss Tropical and Public Health Institute (Swiss TPH) conducted its 37th symposium entitled 'Helminth infection: from transmission to control'. More than 200 participants gathered in Basel, Switzerland during this 2-day event to review, discuss and debate the latest advances and remaining challenges in the epidemiology and control of helminth infections. Some 30 national and international speakers presented their work grouped into six main topic areas: (i) helminth diagnosis; (ii) anthelminthic drugs; (iii) anthelminthic drugs and beyond; (iv) helminth infection: pain and gain; (v) helminth transmission; and (vi) from control to elimination/eradication. The current volume of *Advances in Parasitology* assembles seven authoritative pieces from experts presenting at the Swiss TPH winter symposium in 2017.

Helminth infections have plagued humans and animals since thousands of years (Tayeh et al., 2017) and continue to do so (Collaborators, 2018). Hanspeter Marti from Swiss TPH sets the stage by taking us back into historic times describing fascinating – and at times bewildering – stories of common beliefs built around helminth infections and experiments that ultimately elucidated their life cycles and providing the necessary knowledge base on how to prevent infection, control morbidity and interrupt transmission (Marti, 2019).

Still today, helminth infections are of considerable public health concern, particularly in tropical and subtropical regions of the world where poverty is rife and access to clean water, improved sanitation and hygiene lags behind. Helminth infections can debilitate entire communities affecting the nutritional status and cognitive development or induce tissue damage resulting in blindness, lymphedema and other morbidities exacerbating the vicious cycle of poverty. The review put forth by Charles King from Case Western University in the United States of America provides new insights into the true disease burden of helminth infections and highlights recent successes and remaining challenges in controlling helminth infections (King, 2019).

The third piece by Virak Khieu from the Ministry of Health in Cambodia and colleagues summarizes the situation of liver fluke (i.e. *Opisthorchis viverrini*) infections. Surprisingly little is known on *O. viverrini* in Cambodia, where a dedicated national liver fluke control programme only started a decade ago. It is conceivable that *O. viverrini* infection is underreported in

Cambodia and transmission might rise in the future due to an increase in aquacultural food production, planned construction of a large number of hydropower dams or change of environmental conditions due to climate change, in the absence of adequate sanitary conditions. To tailor setting-specific interventions, well designed epidemiological studies, rigorous surveillance and adequate responses are required (Khieu et al., 2019).

The focus of Chapters 4–6 is on treatment and control of major helminth infections (Ames et al., 2019; Moser et al., 2019; N'Goran et al., 2019). Eliézer N'Goran from the Université Félix Houphouët-Boigny, Abidjan, Côte d'Ivoire and coauthors from the Pediatric Praziquantel Consortium take us onto the research and development journey of new drugs highlighting the challenges observed and lessons learned conducting a phase II study in the western part of Côte d'Ivoire with a pediatric formulation of praziquantel to treat schistosomiasis in preschool-aged children. Oral dispersible tablets should be suitable for use in young children (3 months–6 years), and hence, will provide a much safer and more convenient treatment than crushed praziquantel tables for this age group (N'Goran et al., 2019).

Compelling evidence is provided by the findings from Wendelin Moser and colleagues from Swiss TPH that combination chemotherapy is the way forward for the treatment of soil-transmitted helminth infection. Their network meta-analysis revealed superior efficacy of co-administered treatment over monotherapy, particularly for the whipworm *Trichuris trichiura*. Hence, combination chemotherapy should be considered for the post-2020 strategy for the control and elimination of soil-transmitted helminth infections. However, the availability of effective drugs and treatment regimens will not be sufficient to eliminate helminth infections (Moser et al., 2019).

The article by Heather Melanie R. Ames from Swiss TPH and colleagues reviewed studies on community and drug distributor perceptions and experiences of mass drug administration campaigns for the elimination of lymphatic filariasis. Their review highlights that communities as well as community drug distributors lack knowledge and information about lymphatic filariasis and mass drug administration campaigns, which influences compliance. Moreover, length, timing, level of community and health system involvement, access to care whenever adverse events occur and inadequate numbers of drug distributors influenced treatment coverage (Ames et al., 2019).

By means of a systematic review, Alon Vasiman from the University of Toronto, Canada and coauthors demonstrate the significant technological advances made in the past few years in handheld and mobile phone microscopy, including the CellScope and the Newton Nm1 microscope. The authors also stress that a host of mobile diagnostic tools are currently in the pipeline. Once properly validated and embedded within the health system, these portable and low-cost diagnostic tools might lay the foundation for significant healthcare advances (Vasiman et al., 2019).

The collection of articles in this special issue of *Advances in Parasitology* highlights continued progress in our understanding of the diagnosis, epidemiology, disease burden, treatment and control of helminth infections. Yet, research is still needed and properly validated tools must become accessible by the most marginalized communities in order to move from morbidity control to interruption of helminth transmission.

REFERENCES

Ames, H.M.R., Zuske, M., King, J.D., Steinmann, P., Bosch-Capblanch, X., 2019. Community and drug distributor perceptions and experiences of mass drug administration for the elimination of lymphatic filariasis: a rapid review of qualitative research. Adv. Parasitol. 103, 117–149.

GBD 2017 DALYs, HALE Collaborators, 2018. Global, regional, and national disability-adjusted life-years (DALYs) for 359 diseases and injuries and healthy life expectancy (HALE) for 195 countries and territories, 1990-2017: a systematic analysis for the Global Burden of Disease Study 2017. Lancet 392 (10159), 1859–1922.

Khieu, V., Fürst, T., Miyamoto, K., Yong, T.S., Chai, J.Y., Huy, R., Muth, S., Odermatt, P., 2019. Is *Opisthorchis viverrini* emerging in Cambodia? Adv. Parasitol. 103, 31–73.

King, C.H., 2019. Helminthiasis epidemiology and control: scoring successes and meeting the remaining challenges. Adv. Parasitol. 103, 11–30.

Marti, H., 2019. The discovery of helminth life cycles. Adv. Parasitol. 103, 1–10.

Moser, W., Schindler, C., Keiser, J., 2019. Drug combinations against soil-transmitted helminth infections. Adv. Parasitol. 103, 91–115.

N'Goran, E., Aka, N.A.D., Ouattara, M., Huber, E., Bezuidenhout, D., Kourany-Lefoll, E., 2019. Challenges and lessons from conducting a paediatric clinical trial in Sub-Saharan Africa: the case of the praziquantel oral dispersible tablets Phase II study in Côte d'Ivoire. Adv. Parasitol. 103, 75–89.

Tayeh, A., Cairncross, S., Cox, F.E.G., 2017. Guinea worm: from Robert Leiper to eradication. Parasitology 144 (12), 1643–1648.

Vasiman, A., Stothard, J.R., Bogoch, I.I., 2019. Mobile phone devices and handheld microscopes as diagnostic platforms for malaria and neglected tropical diseases (NTDs) in low-resource settings: a systematic review, historical perspective and future outlook. Adv. Parasitol. 103, 151–173.

CHAPTER ONE

The Discovery of Helminth Life Cycles

Hanspeter Marti[1]
Swiss Tropical and Public Health Institute, Basel, Switzerland; University of Basel, Basel, Switzerland
[1]Corresponding author: E-mail: hanspeter.marti@swisstph.ch

Contents

Abstract

The origin of worms, which were occasionally excreted by some individuals, has left humans wondering for centuries. The story of the gradual discovery of the life cycles of the parasitic helminths is a fascinating one, peppered with misconceptions, misguided by dogmas of the church and enlightened by brave men, who did not recant when they were attacked by the opinion leaders of their time. This article will try to pinpoint some of the key discoveries which turned out to be milestones for the elucidation of the secret life of helminths.

1. THE BEGINNINGS

Parasites have been with us since eons. Some of them were already with us when our ancestors left Africa and set out to conquer the world, others were acquired along the way or a "gift" from our domesticated animals (Morand et al., 2014).

Humans have always been wondering, where the worms come from. The belief in spontaneous generation out of carrion or mud can be traced back to antiquity. Aristotle had seen living worms in Cysts in pig's tongues (Zimmer, 2001). Furthermore, no one had ever seen an *Ascaris* or a tapeworm invading a body, and as these creatures seemed to come from

Advances in Parasitology, Volume 103
ISSN 0065-308X
https://doi.org/10.1016/bs.apar.2019.02.001

nowhere and despite their resemblance to worms did not seem to go anywhere, it was assumed that they were created spontaneously in the bodies. Therefore many scientists concluded that the worms are not causing diseases. This belief persisted for almost 2000 years, until the rise of Christianity when it was clearly opposed by the bible. If all life was created within six days as described in the Genesis, the worms must have also been created during that period and therefore already have been present in paradise. The thought of Adam and Eve being infected with parasitic creatures seemed to be absurd and incompatible with a life in paradise, moreover as they must have passed on these creatures to their offspring, as remarked by the German naturalist Peter Simon Pallas (1741–1811)

> *It is difficult to imagine a paradisiacal happy life, knowing that Adam and Eve must have harboured three different species of tapeworms as well as roundworms and pinworms and passed them on to their offspring*
>
> ***Enigk (1986) and Lieven (2013).***

Yet the theory of the spontaneous generation of life had at time already been challenged by the Florentine doctor Francesco Redi in 1668. He put meat in jars and left some of them uncovered, exposing it to flies. The other jars were covered with gauze, allowing air to diffuse, but flies could not access. As maggots developed only on the meat in uncovered jars, he concluded, after having conducted further experiments and watching the maggots transform into flies: "Omne vivum ex vivo" (All life comes from life) (Gottdenker, 1979).

This, however, did not explain how the worms get into a human being. After the invention of the microscope it was realized, that the male sperm not only consisted of fluid, but contained tiny moving creatures. So up to the beginning of the 19th century, the sperm cells were considered to be parasitic worms, while the fluid was believed to be responsible for the generation of new life (It has to be noted that the ovum was discovered only in 1827) (Baer, 1827). Still, this was not a satisfactory explanation for many scientists. They always found only adult egg laying flukes, but never young, growing ones. Unlike in humans or other mammals, young offspring resembling the adults were lacking.

2. ONE WORM CAN HAVE DIFFERENT SHAPES

In 1831 Karl Mehlis observed that the eggs of certain flukes contain an embryo, which bursts out of the eggshell when put into water and started swimming around vividly. Still, that hairy thing (the miracidium) only faintly resembled an adult fluke. Other observers noted further forms swimming in the water looking much more fluke-like, but having a tail which the adult worm did not. It was Johann Steenstrup, a Danish zoologist, who observed that these forms (the cercaria) shed their tail and penetrated snails. When examining these snails, he found that the cercariae had formed a hard cyst. Furthermore he came to realize that the well-known sack-like animal inside many snails contained creatures which resembled the free swimming forms with the tail. Could it be that all these forms were linked somehow and were just different forms of the same species, although many were then considered to be a species in its own right and had been given a Latin name? He finally put all the pieces of the puzzle together and in his 1842 publication coined the term of "alternation of generations", the succession of sexual and asexual generations (Steenstrup, 1842). Consequently he also suggested that the bladder worms, known today as cysticerci, were stages of so far unknown parasites. Of course his theory did not go unchallenged - Darwin's "Origin of species" would only be published 15 years later — and Steenstrup was blamed to be a heretic as his claim was opposed to the predominant religious beliefs of the time. Yet he had delivered the most convincing argument against the spontaneous generation of vermin, and shortly afterwards Friedrich Küchenmeister would hammer the last nail into the coffin of this theory.

3. THE DISCOVERY OF THE INFECTIVE STAGE

Küchenmeister was a German doctor with a medical practice in Dresden. He was a devout Christian and recognized that Steenstrup's theory not only terminated the heresy of spontaneous creation, but would also solve the riddle of the bladder worms. Considered to be deformed tapeworms gone astray, it would in his opinion contradict the wisdom of the Creator who would not create anything without purpose, not even bladder worms. Building on Steenstrup's suggestion, Küchenmeister realized that mostly bladder worms were found in prey-animals such as mice, rabbits or pigs, while the tapeworms were predominantly found in predators like cats, foxes or humans. Could it be possible that the bladder worms were not a dead end form, but comparable to the cyst Steenstrup had found in the snails after having been penetrated by a cercaria and harboring a tiny fluke? This would possibly signify that the bladder worm was a young stage of a tapeworm,

Figure 1 Japetus Steenstrup. *http://www.kb.dk/images/billed/2010/okt/billeder/object-475639/en/.*

only waiting to be devoured by a predator and then transform into an adult worm. In 1851 Küchenmeister set out to prove his speculation. He fed forty bladder worms recovered from rabbit meat to captured foxes. After a few weeks he was able to recover thirty-five tapeworms. Encouraged by his success, he undertook the same experiment feeding bladder worms recovered from mice to cats, again with best success. His publication in 1852 (Küchenmeister, 1852) was an absolute sensation which stunned the academic world, which was hopping mad that an amateur had solved the mystery they had failed to unravel for such a long time. They tried to discredit his findings, pointing out that the infection experiments did not always work. The fact of host specificity was not yet recognized and so the same experiment e.g. failed, when Küchenmeister fed bladder worms from pigs to dogs. However, realising the frequent occurrence of *Taenia solium* among butchers and kitchen aids in Dresden, he suspected the reason: The bladder worms in pigs had to be eaten by a human in order to grow to an adult tapeworm. The two forms actually were one and the same species. In 1854 he was going to prove his theory. He learned about a convicted murderer in a neighbouring village to be executed within the next few days. As it happened, his wife realized the very same evening that the roast pork they were eating for dinner contained some bladder worms (!) and Küchenmeister immediately rushed to the vendor to beg for some more raw meat from the slaughtered pig. In his later report (Küchenmeister, 1855) he described in great detail how he offered the unsuspecting delinquent spiked noodle soup and blood sausage, where some fat particles had been replaced with cysticerci from the pork meat. The convict enjoyed his meal a lot and even asked for seconds. So he was treated with a total of 75 cysticerci between 72 and 12 h before his execution. At autopsy, although delayed by 48 h, Küchenmeister was able to retrieve 10 tiny *Taenia* worms. Five years later he repeated the experiment, feeding the cysticerci to the convict four months before his execution. This time he was able to retrieve from the intestines a tapeworm more than one m in length. Besides proving Steenstrup and himself right, Küchenmeister had discovered another important fact: parasites did not always have to leave their host and actively seek another one — they could stay put in one host and just wait for being eaten by another one. The best example to demonstrate this was the detection of the life cycle of *Trichinella*.

4. A WORM AT THE CENTRE OF PUBLIC AWARENESS

In 1835 the medical student James Paget discovered a nematode encapsulated in human muscle tissue. His Professor Richard Owen claimed the discovery for himself and named the new parasite *Trichinella spiralis*. He considered the parasite to be harmless, as due to its tiny size (!) it would not even be noticed by an infected human. It was only a quarter of a century later that Friedrich Zenker, a German pathologist from Dresden (again), discovered myriads of *Trichinella* worms during an autopsy of a young maid-servant who allegedly had died of typhoid fever. When Zenker learned that more people from the very same farm had fallen severely ill following the Christmas dinner, he carried out more investigations which finally led to the source of the infection: ham and sausages from a slaughtered pig (Enzensberger, 2001). The whole life cycle was finally deciphered by Rudolph Virchow, famed German pathologist, anthropologist, prehistorian and politician. He examined meat from a fresh hospital ccorpse, which contained numerous white flecks in its muscles, like described earlier by Owen. Virchow indeed could confirm that these flecks were *T. spiralis* larvae and fed the meat to an unsuspecting old dog. The animal succumbed a few days later to the infection and at autopsy he found adult worms in the small intestine and correctly pieced the life cycle together. But to his great merit he did not stop there. Most importantly, he discovered that when infected meat is heated to at least 65 °C for 10 min, the larvae would be killed and the meat was no longer infectious. He began travelling around the countryside and tried to convince the people from the virtues of eating only well-done pork products (Campbell, 1983). This was not an easy task in a country where raw meat products were and still are commonly eaten by virtually the whole population and lightly smoke-cured ham is almost a national speciality. The hostility did not take long to emerge and his public appearances - actually a real public health campaign - often created disbelief and he was contradicted when explaining the way of infection with the tiny parasite. As the story goes, one of the disbelievers was out to prove the incorrectness of the theory by eating a heavily infected sausage and paid for it with his life (Billings, 1884).

After large outbreaks in 1863 and 1865 in the German towns of Hettstedt and Hedersleden, involving several hundred persons with a case fatality rate of up to 30% (Campbell, 1983), Virchow was masterminding trichinoscopy as a way to eliminate infected pork as a source of infection. It finally became compulsory in the kingdom of Prussia in 1877 and in the whole of Germany in 1900 (Hinz, 1991) and was consequently enforced. So in 1895 in Prussia alone there were 27089 meat inspectors for trichinellosis, a number that almost matched the number of the US Army of that time (Enzensberger, 2001)! The enormous effort paid off. According to estimates there were about 15000 cases diagnosed in Germany in 1900. Within 50

Figure 2 Rudolf Virchow. *http://resource.nlm.nih.gov/101431213.*

years this number dropped dramatically. In the first ten years of the new Millennium there were only four positive pigs identified among the roughly 453 million domestic pigs examined, while there were 92 positive cases among wild boars (RKI, 2018). In the US, where pork was not subjected to mandatory trichinoscopy, there were estimates that as late as 1940 there were about 21 million Americans who had been infected with trichinellosis during their lifetime (Stoll, 1947).

5. THE DETECTION OF THE INTERMEDIATE HOST

The last big missing puzzle in the life cycles of helminths was identified by Patrick Manson, later knighted and named the "Father of Tropical Medicine". When working as a young physician in the port of Amoy in southern China, he was struck by the view of patients suffering from elephantiasis. Then he heard of a discovery by T.R. Lewis in India, that microfilaria were associated with this symptom (Eldridge, 1992). By very careful observations in patients with elephantiasis he discovered that sheathed microfilaria were present in great abundance in the blood during the night with a peak around midnight, but were present at much lower rates or even absent during the day. Furthermore, the microfilaria did not develop any further in their human host. Just as Steenstrup had assumed some 30 years earlier, Manson suspected that the microfilaria were only the young forms of some adult worm, and as never intermediate forms had been found, there must be some other animal serving as intermediate host. Combining the evidence, he suspected that mosquitoes, which bite at nighttime, might be the missing link he was looking for. He built a small house and had people suffering from elephantiasis, including his gardener, sleep in there. On the following morning, he trapped the mosquitoes that had taken a blood meal and dissected them. He found that the microfilaria were not killed by the digestive juices, but on the contrary were very active. Manson could demonstrate that inside the mosquito the larvae shed their sheath and moved from the alimentary tract to the thoracic muscles, where they underwent further development (To and Yuen, 2012). The publication of his observations in 1877 in China and 1878 in the Journal of the Linnean Society Manson (1878) was a medical breakthrough, although his first conclusions were not correct. At that time little was known about the life of mosquitoes. It was believed that they lived only a short period of time and died shortly after oviposition on a water surface. Therefore Manson concluded that after

Figure 3 Patrick Manson. *https://wellcomecollection.org/works/ndacujyq.*

the death of the insect the larvae would escape into the water and humans would become infected when drinking contaminated water (Eldridge, 1992). It was only more than 20 years later in 1899 when Thomas Bancroft demonstrated that infected mosquitoes can inoculate the larvae into humans by biting. This does, however, in no way dwarf Manson's monumental discovery of the role of the tiny bloodsuckers in the human to human transmission of parasites. These observations gave rise to the theory, that also the scourge of malaria could be transmitted by mosquitoes, which was finally proven in 1898 through the work of Ronald Ross in India, demonstrating the full life-cycle of avian malaria in mosquitoes (Ross, 1897).

Figure 4 Thomas Lane Bancroft. *http://rosettadel.slq.qld.gov.au/delivery/DeliveryManagerServlet?change_lng=en&dps_pid=IE401975.*

Nowadays it is known, that humans can be infected with nearly 300 different species of worms (Cox, 2002), which is about a tenfold increase in numbers since the beginning of the 20th century. Many of these infections are very rare or just accidental, and every now and then a new pest is described (Bradbury et al., 2018). Yet without the pioneering works of the great names in the field of parasitology, elucidation the fundamentals of this field often in a gruesome way, we would still be wondering about these infections and not at all be able to set targets by which year some of them should be eradicated or at least eliminated as a public health problem.

REFERENCES

Baer, K. E. v., 1827. De ovi mammalium et hominis genesi, Leipzig. Lipsiae Sumptibus L. Vossii, Leipzig.

Billings, F.S., 1884. The Relation of Animal Diseases to the Publiv Health and Their Prevention. Appleton, New York, p. 17.

Bradbury, R.S., Breen, K.V., Bonura, E.M., Hoyt, J.W., Bishop, H.S., 2018. Case report: conjunctival infestation with *Thelazia gulosa*: a novel agent of human Thelaziasis in the United States. Am. J. Trop. Med. Hyg. 98 (4), 1171–1174.

Campbell, W.C., 1983. Trichinella and Trichinosis. Plenum Press, New York.

Cox, F.E., 2002. History of human parasitology. Clin. Microbiol. Rev. 15 (4), 595–612.

Eldridge, B.F., 1992. Patrick Manson and the discovery age of vector biology. J. Am. Mosq. Contr. Assoc. 8 (3), 215–220.

Enigk, K., 1986. Geschichte der Helminthologie im deutschsprachigen Raum. Gustav Fischer Verlag (quoted by A. Fürst von Lieven).

Enzensberger, U., 2001. Parasiten. Ein Sachbuch. Eichborn Verlag.

Gottdenker, P., 1979. Francesco Redi and the fly experiments. Bull. Hist. Med. 53 (4), 575–592.

Hinz, E., 1991. Trichinellosis and trichinellosis control in Germany. Southeast Asian J. Trop. Med. Publ. Health 22 (Suppl), 329–333.

Küchenmeister, F., 1852. Ueber die Umwandlung der Finnen in Bandwürmer. Vierteljahr Schr Prakt Heilk Prag 33, 106–158.

Küchenmeister, F., 1855. Experimenteller Nachweis, dass *Cysticercus cellulosae* innerhalb des menschlichen Darmkanales sich in *Taenia solium* umwandelt. Allgemeine deutsche Naturhistorische Zeitung 1, 186–191.

Lieven, A. v., 2013. Die Vielfalt der Fadenwürmer. Naturwiss. Rundsch. 66 (4), 173–183.

Manson, P., 1878. On the development of *Filaria sanguinis hominis*, and on the mosquito considered as a nurse. Zool. J. Linn. Soc. 14 (75), 304–311.

Morand, S., McIntyre, K.M., Baylis, M., 2014. Domesticated animals and human infectious diseases of zoonotic origins: domestication time matters. Infect. Genet. Evol. 24, 76–81.

RKI, 2018. Epidemiologisches Bulletin des Robert-Koch-Instituts 3.

Ross, R., 1897. On some peculiar pigmented cells found in two mosquitos fed on malarial blood. Br. Med. J. 2, 1786–1788.

Steenstrup, J.J., 1842. Ueber den Generationswechsel, oder die Fortpflanzung und Entwickelung durch abwechselnde Generationen, eine eigenthümliche Form der Brutpflege in den niedrigen Thierclassen. Reitzel, Copenhagen.

Stoll, N.R., 1947. This wormy world. J. Parasitol. 33 (1), 1–18.

To, K.K., Yuen, K.Y., 2012. In memory of Patrick Manson, founding father of tropical medicine and the discovery of vector-borne infections. Emerg. Microb. Infect. 1 (10), e31.

Zimmer, C., 2001. Parasite Rex: Inside the Bizarre World of Nature's Most Dangerous Creatures. Touchstone.

CHAPTER TWO

Helminthiasis Epidemiology and Control: Scoring Successes and Meeting the Remaining Challenges

Charles H. King[*,§]

[*]Center for Global Health and Diseases, WHO Collaborating Centre for Research and Training on Schistosomiasis Elimination, Case Western Reserve University School of Medicine, Cleveland, OH, United States

[§]Schistosomiasis Consortium for Operational Research and Evaluation (SCORE), University of Georgia, Athens, GA, United States

E-mail: chk@cwru.edu

Contents

Abstract

Parasitic helminth infections remain a significant challenge to global health. These are highly prevalent diseases, affecting over 1 billion persons worldwide. Their prevalence is closely linked to the presence of severe poverty and its associated sub-standard housing and sanitation. The last decade has seen a remarkable increase in our understanding of the true disease burden of helminth infections, and there has been increasing momentum on the part of national and non-governmental

Advances in Parasitology, Volume 103
ISSN 0065-308X
https://doi.org/10.1016/bs.apar.2018.08.001

developmental organizations for prevention and control of these diseases. The expansion in mass treatment programmes for their control has yielded some significant successes. However, challenges remain in terms of ecological heterogeneity in transmission, incomplete drug uptake, and the likelihood of emerging drug resistance. The development of new, more-sensitive diagnostics is now broadening our knowledge of infection prevalence and of the risk of reinfection and has enhanced our knowledge of the prevalence of concurrent helminth infections. Adoption of these new diagnostic techniques for large-scale screening and surveillance will require adaptation of current mass treatment guidelines for control as programmes move from initial morbidity control objectives toward coordinated interventions aimed at local elimination.

1. WHAT ARE THE HELMINTHS?

Helminth parasites constitute a health burden of major proportion—over one billion people worldwide carry these long-lived infections and suffer daily from the consequences of infection-associated pathologies (Maguire, 2015; GBD, 2016 Disease and Injury Incidence and Prevalence Collaborators, 2017). Helminths are multicellular animals adapted to living most of their lives within vertebrate hosts. They are substantially larger in size (0.5 cm to 25 m) than other infectious pathogens, and their interaction with their human hosts is considerably more complex (Lustigman et al., 2012). Each helminth's life cycle involves transition from egg into intermediate larva, and then into adult stages, but most helminth parasites cannot reproduce within the human host. Helminth species require a stage of development either in soil, in water, or in an intermediate host, and establishment of human infection requires either ingestion of helminth eggs or larvae, skin contact and penetration by infectious larvae, or skin inoculation via insect vectors. For this reason, the intensity of a patient's infection and the severity of his or her infection-related morbidity are functions of continuing exposure to contaminated environments, which allows progressive accumulation of more intense worm burden (Maguire, 2015). Where environmental factors favour transmission, as is often the case in humid tropical locations, then high intensity infections are much more common and polyparasitism (infection with more than one species of parasite) is frequent (Bisanzio et al., 2014).

Helminth parasite species are members of two phyla within the animal kingdom: Nematoda (roundworms) and Platyhelminthes (flatworms). The platyhelminth parasites come from two classes, Trematoda and Cestoda (Fig. 1). The trematodes are commonly referred to as flukes, whereas mature

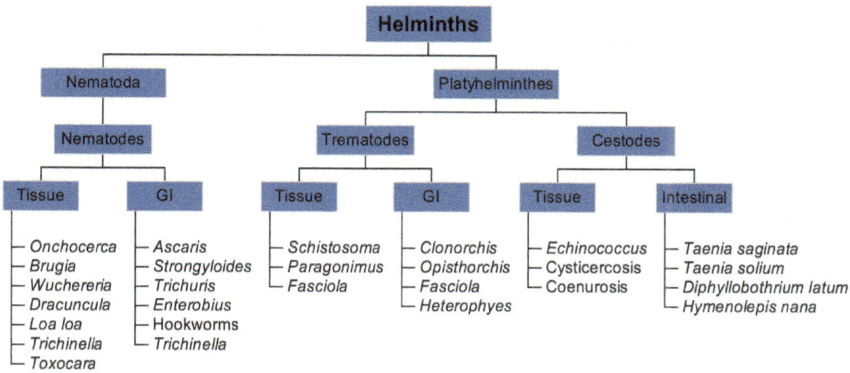

Figure 1 Diagram of the phyla, classes, and genera for the most prevalent helminth infections of humans. Adult helminths may either colonize the gastrointestinal tract or liver (designated 'GI'), or colonize skin, the lymphatic system, the bloodstream, or other body organs (designated 'Tissue').

cestode parasites are most often referred to as tapeworms. When humans serve instead as intermediate hosts for cestodes, these infections are referred to as cysticercosis (*Taenia solium*), hydatidosis (*Echinococcus* spp.), and coenurosis (caused by canine *Taenia* spp.) (King and Fairley, 2015). For many helminths, utilization of multiple hosts (intermediate and definitive), long adult lifespans (years), and abundant egg production allow species reservoirs to persist locally during periods of adverse weather (drought, winter), then immediately restart the process of transmission once favourable conditions resume.

2. GLOBAL BURDEN OF HELMINTH INFECTIONS

Helminth infections affect billions of people around the world. The latest 2016 prevalence estimates of the Global Burden of Disease (GBD) Project (GBD, 2016 Disease and Injury Incidence and Prevalence Collaborators, 2017) indicate there are approximately 800 million cases of ascariasis, 450 million cases of hookworm, 435 million cases of trichuriasis, 190 million cases of schistosomiasis, and 75 million cases of food-borne trematodiasis worldwide. Lymphatic filariasis (29 million cases) and onchocerciasis (15 million) are now less common due to intensive control efforts implemented over the last three to four decades. Newly completed GBD estimates for the prevalence of neurocysticercosis (NCC, 2.7 million) and cystic echinococcosis (CE, 1 million) now reveal the significant burden

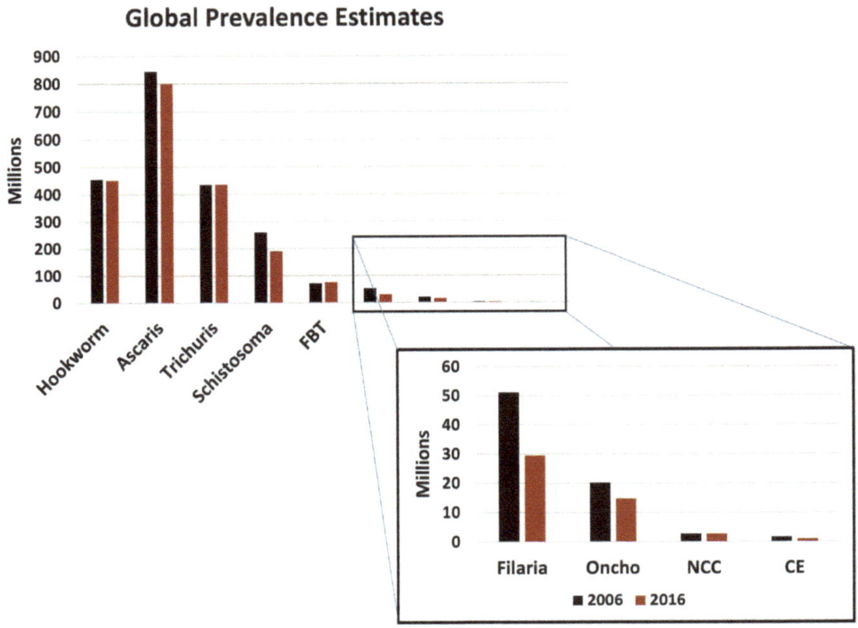

Figure 2 Global prevalence estimates for the world's most common helminth infections. Darker bars indicate prevalence estimates for 2006, and lighter bars indicate estimates for 2016. Notable reductions were observed in the estimates for lymphatic filariasis prevalence (Filaria, 42% reduction), cystic echinococcosis prevalence (CE, 39% reduction), onchocerciasis prevalence (Oncho, 27% reduction), and *Schistosoma* prevalence (27% reduction) across this ten year span. See (GBD, 2016 Disease and Injury Incidence and Prevalence Collaborators, 2017) for details.

of these seriously disabling diseases as well. Fig. 2 indicates the GBD estimated prevalences of each infection in 2006 as compared to 2016. Remarkable interval changes were: a 47% reduction in lymphatic filariasis, a 39% reduction in CE, a 27% reduction in schistosomiasis, and a 27% reduction in onchocerciasis during the intervening decade. Unfortunately, there was very little change in the prevalence estimates for the soil-transmitted helminths, *Ascaris lumbricoides*, hookworm, and *Trichuris trichiura* (STH) during that period.

2.1 Who Gets Helminths?

When we examine the map of where helminth infections are most abundant (Fig. 3), it is apparent that there is a greater concentration of cases within the tropical and sub-tropical regions. This finding is in part related to local climate factors that facilitate the development of STH eggs and/or larvae

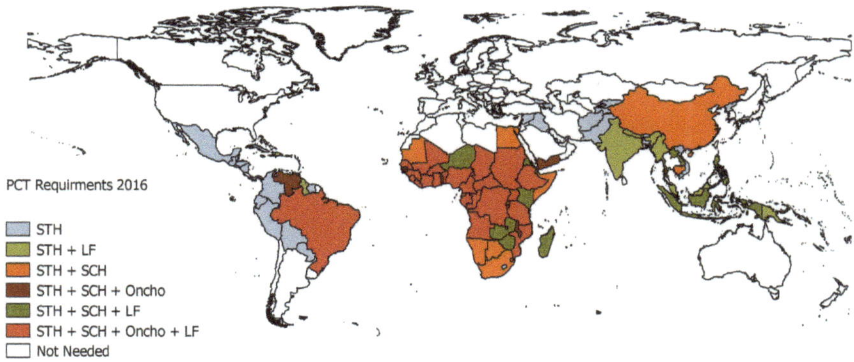

Figure 3 World map showing the distribution of helminth burden in terms of requirements for preventive chemotherapy (PCT) intervention in 2016. After disease mapping surveys, the shaded countries were determined to require PCT for soil-transmitted helminths (STH), lymphatic filariasis (LF), onchocerciasis (Oncho), and/or *Schistosoma* infections (SCH), either singly or in combination. *Source: WHO PCT databank WHO, 2018. Preventive Chemotherapy Databank [Online]. Available: http://www.who.int/neglected_diseases/preventive_chemotherapy/databank/en/index.html.*

in the environment, and other factors that foster the constant or frequent presence of transmitting insect and snail intermediate host species.

Nevertheless, helminth infections can also be transmitted outside of these zones, including in the sub-arctic and arctic regions. In all parts of the world, household poverty and lack of adequate sanitation remain powerful contributing factors for infection risk. There is a striking similarity between the maps of helminthic disease prevalence (Fig. 3) and of national prevalence of severe poverty (Fig. 4), defined as subsisting on less than \$1.90 US dollar

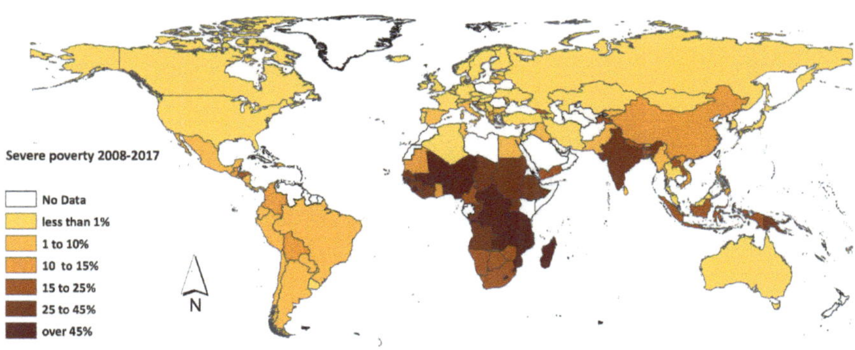

Figure 4 World map of the prevalence of severe poverty (defined as percent of population living on $<$ \$1.90 US dollar per day) between 2008 and 2017, according to the World Bank data (http://databank.worldbank.org/data/home.aspx). Note the very similar distribution of severe poverty compared with the frequency and multiplicity of helminth infections found in Fig. 3.

equivalent purchasing power per day (http://databank.worldbank.org/data/home.aspx). Thus, the helminth infections are not strictly 'tropical' diseases. They can more rightly be viewed as global diseases of poverty, for which sustainable development will be ultimately needed to bring them under full control. Moves to achieve the United Nation's 2030 Sustainable Development Goals (SDG), numbers 1 (poverty reduction), 3 (good health and well-being) and 6 (clean water and sanitation)), will significantly contribute to achieving permanent helminthic disease elimination (United Nations General Assembly, 2015).

3. A NEW APPRECIATION OF THE FUNCTIONAL IMPACT OF HELMINTH INFECTIONS

In past policy deliberations, low-to-moderate intensity helminth infections were often described as 'asymptomatic'. In order to maximize the implementation of available resources, rationing of drug delivery was suggested, in which priority for treatment was given only to reducing the prevalence of heavy infections, which were assumed to be a proxy for infection-associated disease risk (Warren, 1982). This proved to be a false assumption, in that persons with chronic infections still manifest functional symptoms and remain at continuing risk for acquiring increased worm burden and more acute or permanent complications of their continuing infection (King, 2015). In the 1990s, as public private partnerships emerged for the control of onchocerciasis and LF, greater availability of antihelminthic drugs such as ivermectin, diethylcarbamazine (DEC), and albendazole changed the focus of intervention toward 'preventive chemotherapy' (PCT). This approach, based on annual delivery of antihelminthic drug treatment, repeatedly reduces parasite burden for target populations, with the aim of limiting the cumulative damage of recurrent infections (Lustigman et al., 2012; WHO, 2006). Establishment of the Schistosomiasis Control Initiative in the early 2000s led to implementation of many large-scale national programmes for population-based annual delivery of praziquantel, often combined with concurrent treatment for STH. With increased demand for praziquantel, manufacturers of generic versions of the drug entered the market, which resulted in reduction in drug costs by more than 90% (Fenwick et al., 2009). Economies-of-scope were realized by combining delivery of PCT for other helminth infections during the same control team visit (Leslie et al., 2013).

It was fortunate that the scope of the sub-populations targeted for remedial treatment and for mass treatment was broadened in the PCT recommendations. An increasing number of studies have demonstrated there are 'subtle' or functional morbidities caused by helminth infections that have a significant impact on personal performance and on health-related quality-of-life. These are not highly symptomatic or overt conditions that will drive people to seek immediate medical care, but they nevertheless contribute substantially to poor human performance and the cumulative burden of helminth-related disease. Randomized trials and observational studies of treatment outcomes indicate the association of STH and schistosomiasis with anaemia, chronic pain, growth stunting, poor school performance, and cognitive deficits (King et al., 2005; Ezeamama et al., 2018; Pabalan et al., 2018; Musuva et al., 2017). Depression and other psychological trauma also occur as a consequence of social stigmatization, for example, when lymphedema causes disfigurement of the limbs or genitals, or when urogenital schistosomiasis contributes to secondary infertility (Hofstraat and Van Brakel, 2016; Miller-Fellows et al., 2017). Inclusion of these under-recognized long-term pathologies in the scope of infection-associated diseases indicates the importance of including treatment of pre-school age children and of young adults in future mass-treatment campaigns.

4. SUCCESSES IN HELMINTH CONTROL

Control of helminth infections, which represent a large proportion of the 'neglected' tropical diseases (Hotez et al., 2008), has re-emerged as a public health priority in the last decade. Development of broad based mass-treatment strategies (Molyneux et al., 2005), and their successful implementation on national scales, have contributed to substantial reductions in cases of helminth infections (Fig. 2) and to reductions in the prevalence of their associated morbidity. Intensive collaboration between WHO and donor partners resulted in the creation of a 'roadmap' document for implementation through 2020 (Savioli and Daumiere, 2012) and joint issuance of the 2012 London Declaration for Neglected Tropical Diseases (http://unitingtocombatntds.org/london-declaration-neglected-tropical-diseases). PCT has been particularly emphasized for LF, onchocerciasis, schistosomiasis, and STH infections, along with i) intensified disease management to reduce current morbidities, ii) vector and intermediate host control, iii) combined public health and veterinary public health control of zoonotic sources, and iv) provision of safe water, sanitation and hygiene (Savioli and Daumiere, 2012).

A key feature of successful implementation was the integration of PCT to provide concurrent delivery of two or more antihelminthics in areas affected by more than one helminth parasite (WHO, 2006). Creation of a combined drug donation request system facilitated the integration of national NTD control projects, and has also enhanced data flow from endemic areas to national and international planning partners (WHO, 2018). Gap assessments and mobilization of financial support from development programmes has aided political will to achieve control targets, and programme uptake has been successful in most locations. By 2015, over 1.5 billion treatments were being administered to over 1 billion individuals for one or more of the PCT targeted infections: LF, onchocerciasis, schistosomiasis, STH, and trachoma (WHO, 2017a).

For some insect-borne helminth infections such as LF and onchocerciasis, successful implementation of community-wide annual treatments has led to a marked reduction of local transmission to the point where mass treatment can be stopped (Nicholls et al., 2018; Rao et al., 2017; WHO, 2017a). It should be noted, however, that the suspension of mass treatment by these campaigns still requires an aggressive follow-up surveillance system to assure that undetected low-level transmission will not result in re-emergence of significant parasite prevalence and related disease.

5. CHALLENGES IN HELMINTH CONTROL

Ecological analysis of the helminth elimination strategies suggests that there can be a "transmission breakpoint" at low prevalence, below which transmission will falter and human cases will go to zero. The premise rests on the fact that many helminths require sexual reproduction during their adult stages, such that at very low landscape densities, their chances of successful mating are almost zero (MacDonald, 1965; May, 1977). While this theory may apply to some settings, empiric evidence from large-scale programmes indicates that transmission can persist, even when current target treatment coverage and frequency goals are met (Wiegand et al., 2017; Eisenbarth et al., 2016; Rao et al., 2017).

The drive toward total transmission elimination is threatened by a number of factors that defy the assumptions built into mathematical models, and these cause observed programme outcomes to deviate from model projections (Gurarie and King, 2014). Where environmental and human behavioural factors favour transmission, parasite persistence should not be surprising. STH, in particular *A. lumbricoides*, have a very high level of production of infective eggs, and infections are prone to return within a 12 month period (Jia et al., 2012). Heterogeneities in human exposure and susceptibility to infection mean that worms can cluster in specific persons and their preferred locations. A non-random distribution of infection or treatment uptake may then result in a 'superspreader' phenomenon, which causes persistent hot spots of transmission to remain despite a standardized MDA application with high coverage (Woolhouse et al., 1998). Coping with these hot spots is now an essential task that will require design and implementation of adaptive strategies using supplemental control techniques beyond MDA (e.g., vector control, behaviour change, sanitation) to block transmission more effectively and eliminate disease (Lo et al., 2018; Njenga et al., 2011; Booth and Clements, 2018).

Of note, most antihelminthic drugs are not completely curative, i.e., they do not eliminate all infecting worms with a single treatment. Although repeated and more frequent treatments can further lower local prevalence of worm infections, the worms' relatively long lifespans contribute to local persistence of helminth species and the risk of re-emergence once MDA is suspended. The tendency of some individuals to repeatedly miss or refuse MDA (systematic non-compliance) also serves to perpetuate transmission (Farrell and Anderson, 2018). Infected groups that have often been intentionally excluded from PCT or MDA include pre-school age children (because of fears about safe pill delivery and tolerability (Bustinduy et al., 2016)), and women of reproductive age (who have been excluded because of fears about possible drug effects during pregnancy (Theobald et al., 2017; WHO, 2003)). Analysis of transmission dynamics in high risk areas suggest that both groups contribute significantly to the persistence of transmission in treated communities (Toor et al., 2018).

Other continuing challenges in designing and supporting NTD control are i) how to properly address the problem of effectively treating multiple concurrent parasitic infections (polyparasitism), and ii) how to address the issue of controlling helminth infections in zoonotic reservoirs, so as to curtail local risk for infection among humans (Sripa et al., 2017; Budke, 2006). An urgent issue regarding helminth co-infections is the co-endemicity of *Loa loa* and *Onchocerca* filarial species in parts of central and West Africa. Standard MDA with ivermectin given for onchocerciasis can cause severe complications when patients also carry heavy loiasis (Gardon et al., 1997). Recent advances in point-of care screening for circulating *Loa loa* larvae with the use of smartphone microscopy may help to limit risk through implementation of a 'test and not treat' strategy that will prevent ivermectin delivery to persons with high-risk *Loa loa* infections (Kamgno et al., 2017).

6. ON THE HORIZON

What new changes might we expect over the next decade?

6.1 Changes in Parasite Transmission

The ecology of vector species and intermediate host species is likely to change as global climate changes and other anthropogenic landscape changes take place (Zhou et al., 2008; Garchitorena et al., 2017; Cable et al., 2017). In some cases, this will result in expansion of the range and seasonal abundance of helminth-transmitting hosts. However, climate change may also serve to limit their presence within specific regions (McCreesh and Booth, 2013; Cable et al., 2017), or it may serve to facilitate the emergence of competitor/predator species that will reduce the intermediate hosts' presence and so drive the parasites to local extinction (Ivy et al., 2018; Savaya Alkalay et al., 2014). However, increasing human host population density and movement play major roles in the success of pathogen transmission (McCallum et al., 2017). Greater concentration and mixing of susceptible human hosts may serve to amplify rates of exposure to new infection, thus perpetuating transmission even in the face of intensive MDA control efforts.

6.2 Drug Resistance

For the near future, the parasites themselves will probably remain mostly responsive to treatment, although with recent widespread expansion of mass treatment programmes for parasite control, there is very real concern about the potential for emergence of anthelmintic drug resistance over time (Albonico et al., 2015; Crellen et al., 2016). Significant emergence of resistance-associated drug treatment failure has already been experienced in veterinary practice (Diawara et al., 2013; Vercruysse et al., 2018). 'Pharmacovigilance' for continued drug efficacy issue must necessarily remain a priority issue in the realm of helminthic disease control.

6.3 Drug Development and Delivery

The knowledge that many of our anthelmintic drugs are not fully curative, combined with the spectre of potential drug resistance, have stimulated research on the discovery of new anti-parasite compounds. It has also focused interest on the redesign or repurposing of older compounds, to be given either alone or in combination therapy (Moser et al., 2017; Taylor et al., 2015). One of the significant hurdles to achieving population-based worm control has been the inability (or unwillingness) of affected persons to take the recommended medication provided (Krentel et al., 2013; Burnim et al., 2017). Repeated non-compliance with drug treatment creates a reservoir of infection within local populations that can significantly affect the ability of MDA to limit community transmission (Farrell and Anderson, 2018). Significant efforts have been made to improve information exchange with target treatment groups in order to raise their level of awareness of the disease and its necessary treatment(s) and their potential side effects. This helps to prevent participation gaps created by rumours about the occurrence of adverse events with treatment. Paediatric preschool populations have typically been excluded from MDA because of safety concerns about their ability to swallow pills and lack of knowledge on correct dosing and the tolerability of the drugs. For the anti-trematode and anti-tapeworm drug praziquantel, a new, better tasting orally dispersible tablet is under development, which will mean access to treatment for a large sector of at-risk populations not currently receiving treatment for these chronic parasitic infections (Paediatric Praziquantel Consortium [Online]. Available: https://www.pediatricpraziquantelconsortium.org [Accessed 12 April 2018]).

Drug combination therapy may have the potential to provide improved curative or suppressive effects for helminth infections. For example, the use of combined ivermectin, diethylcarbamazine, and albendazole (IDA) for MDA-based control of lymphatic filariasis has been recently shown to provide greater and much longer suppression of microfilaraemia among affected populations (Fischer et al., 2017). This markedly better parasitological response, combined with safety evidence from large scale randomized trials of IDA delivery in five nations, has resulted in revised WHO recommendations for drug-based filariasis control that include IDA as an alternative to previous regimens, particularly where control has proven difficult to achieve (WHO, 2017b).

6.4 New Diagnostics

Over the last two decades, advances in parasite proteomics and genomics have given us the ability to use nucleic acid amplification testing systems, such as PCR and LAMP, for amplification of specific nucleic acid sequences, providing much more sensitive detection of active helminth infections. Use of purified or recombinant antigen fragments now also provide much more specific detection of acquired anti-parasite antibodies, which documents current or past exposure to helminthic parasites. In addition, use of point-of-care parasite antigen detection techniques, based on highly specific monoclonal antibodies, now offers more rapid and precise mapping tools for establishing prevalence of ongoing infections (Colley et al., 2013). Efficient multiplexing of nucleic acid amplification testing (Kaisar et al., 2017), or of serologic testing (Arnold et al., 2017), is delivering broader and more detailed information on the risk for multiple concurrent parasite infections among susceptible populations, and can be used to track the successes (and failures) of implemented combined control efforts.

6.4.1 New Definitions of Infection and Disease

Implementation of the more sensitive diagnostic tests described above has revealed (or has confirmed) that the standard egg-detection microscopy techniques used for detection of most helminthic infections are insensitive for diagnosis of low-level or very light intensity infections. The identification of 'egg-negative/antigen-positive' (or 'PCR-positive') cases often reveals that true prevalence of parasite infection is much higher than what was appreciated in past disease-burden assessments (Colley et al., 2017). It also frequently indicates ongoing helminth transmission where standard (less sensitive) tests have suggested near elimination of infection (Ortu et al., 2017; Haggag et al., 2017).

This raises the urgent need to recalibrate our policy targets for infection-related morbidity control and for parasite elimination goals. While it is true that some parasite-associated pathologies are related to the *intensity* of a helminthic infection, some morbidities are related merely to the *presence* of infection (King, 2015). As such, 'occult' egg-negative infections represent an ongoing health threat to affected individuals, and should still be treated.

In many cases, after implementation of community-based treatment programmes, most of the 'classic' morbidities of helminth infections (as described in textbooks) will begin to disappear from infected areas. However, persistent transmission and frequent re-infection lead to persistence of more subtle, functional morbidities that affect childhood growth and development, haemoglobin levels (Leenstra et al., 2006), and school performance (Pabalan et al., 2018; Ezeamama et al., 2018). Of note, a dozen randomized placebo-controlled trials for schistosomiasis have shown that these 'subtle' forms of morbidity can be significantly improved (within 1–12 months) following treatment with parasite-specific drugs (Fig. 5). In contrast to the current Disability-Adjusted Life Year (DALY) estimates provided by the GBD Project (GBD, 2016 Disease and Injury Incidence and Prevalence Collaborators, 2017), if we now include these infection-related conditions in a more comprehensive assessment of the impact of NTD infections on patient health status, we will achieve a more realistic estimate of their overall impact on human quality of life. This will certainly provide a better estimate of their contribution to the global burden of disease.

6.5 The Movement toward Elimination

Recognition of the disease impact of persistent or recurrent infection, noted above, and of the ecological resilience of helminth transmission even in the face of aggressive mass treatment strategies, leads us to a re-examination of

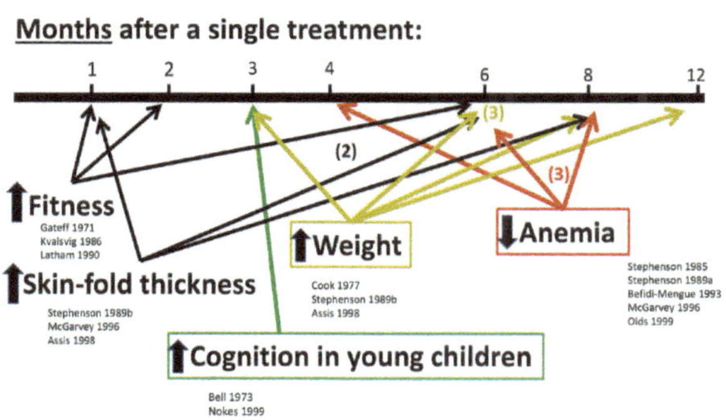

Figure 5 Health benefits identified in twelve placebo-controlled, randomized trials of drug treatment for 'subtle' forms of schistosomiasis-related morbidity. The time line indicates effects observed at follow-up testing performed at different intervals (in months) after therapy. Names and dates below each health outcome refer to the relevant published trials reporting on that outcome (Gateff et al., 1971; Bell et al., 1973; Cook et al., 1977; Stephenson et al., 1985, 1989a, 1989b; Kvalsvig, 1986; Latham et al., 1990; McGarvey et al., 1996; Assis et al., 1998; Nokes et al., 1999; Olds et al., 1999). Thin arrows indicate the time of observation; thick arrows indicate the direction of change in morbidity status after therapy. Numbers in parentheses next to thin arrow indicate more than one study reporting on a given outcome at a specific time interval.

our current guidelines for control. Persistence of transmission and risk of prevalence rebound need to be prevented to truly prevent disease overall, and the best means to achieve this will be local elimination of infections. Elimination, where possible, has been now made a definite goal for schistosomes, STH, and liver flukes in different spheres of public health control (Lustigman et al., 2012), but this will require implementation of more complex intervention strategies to achieve success.

7. SUMMARY

Growing recognition of the health impact of helminth infections has led to a broad-based expansion in research and implementation programmes to limit and prevent these 'neglected tropical diseases' (Molyneux et al., 2005). Table 1 summarizes many of their current successes and the

Table 1 Scoring the Successes and Remaining Challenges for Helminth Control

Successes	Challenges
Wide implementation of PCT for school age children	PCT may not reduce transmission and cannot prevent reinfection
Additional focus on pre-school age children and women of reproductive age	Coverage of pre-school age children and adult women is beyond the scope of current PCT guidelines for some NTDs
Achieving transmission control for LF and *Onchocerca* in some areas. Approval of a new, highly effective triple drug regimen for LF	Remaining hot spots of persistent transmission; need for control of *Loa loa* and for safe control of LF and *Onchocerca* in *Loa loa*-endemic zones
Recognition of the global disease burden of foodborne trematodes, cysticercosis, and echinococcosis	These are multiple species having complex multi-host life cycles that require more complex intervention strategies to obtain control
New, more sensitive and specific diagnostics	Need to align prevalence findings of the new tests with older methods, or we need to completely rethink population treatment thresholds
New drugs, drug combinations, and delivery methods	These new drugs will not be immediately available for PCT/MDA until population safety profiles are established and donations will need to be discussed with the producers
We're now talking about elimination	The last mile will be the hardest part of the job

Abbreviations: *LF*, Lymphatic Filariasis; *MDA*, Mass Drug Administration; *NTD*, neglected tropical disease; *PCT*, Preventive Chemotherapy.

remaining challenges in bringing these complex diseases under full control. Continuing momentum in basic and operational research, combined with very essential political will to achieve control, will ultimately be needed to reach these goals.

REFERENCES

Albonico, M., Levecke, B., Loverde, P.T., Montresor, A., Prichard, R., Vercruysse, J., Webster, J.P., 2015. Monitoring the efficacy of drugs for neglected tropical diseases controlled by preventive chemotherapy. J. Glob. Antimicrob. Resist. 3, 229–236.

Arnold, B.F., Van Der Laan, M.J., Hubbard, A.E., Steel, C., Kubofcik, J., Hamlin, K.L., Moss, D.M., Nutman, T.B., Priest, J.W., Lammie, P.J., 2017. Measuring changes in transmission of neglected tropical diseases, malaria, and enteric pathogens from quantitative antibody levels. PLoS Negl. Trop. Dis. 11 e0005616.

Assis, A.M., Barreto, M.L., Prado, M.S., Reis, M.G., Parraga, I.M., Blanton, R.E., 1998. *Schistosoma mansoni* infection and nutritional status in schoolchildren: a randomized, double-blind trial in northeastern Brazil. Am. J. Clin. Nutr. 68, 1247–1253.

Bell, R.M., Daly, J., Kanengoni, E., Jones, J.J., 1973. The effects of endemic schistosomiasis and of hycanthone on the mental ability of African school children. Trans. R. Soc. Trop. Med. Hyg. 67, 694–701.

Bisanzio, D., Mutuku, F., Bustinduy, A.L., Mungai, P.L., Muchiri, E.M., King, C.H., Kitron, U., 2014. Cross-sectional study of the burden of vector-borne and soil-transmitted polyparasitism in rural communities of Coast Province, Kenya. PLoS Negl. Trop. Dis. 8 e2992.

Booth, M., Clements, A., 2018. Neglected Tropical Disease control - the case for adaptive, location-specific solutions. Trends Parasitol. 34, 272–282.

Budke, C.M., 2006. Global socioeconomic impact of cystic echinococcosis. Emerg. Infect. Dis. 12, 296–303.

Burnim, M., Ivy, J.A., King, C.H., 2017. Systematic review of community-based, school-based, and combined delivery modes for reaching school-aged children in mass drug administration programs for schistosomiasis. PLoS Negl. Trop. Dis. 11 e0006043.

Bustinduy, A.L., Friedman, J.F., Kjetland, E.F., Ezeamama, A.E., Kabatereine, N.B., Stothard, J.R., King, C.H., 2016. Expanding praziquantel (PZQ) access beyond Mass Drug Administration programs: paving a way forward for a pediatric PZQ formulation for schistosomiasis. PLoS Negl. Trop. Dis. 10 e0004946.

Cable, J., Barber, I., Boag, B., Ellison, A.R., Morgan, E.R., Murray, K., Pascoe, E.L., Sait, S.M., Wilson, A.J., Booth, M., 2017. Global change, parasite transmission and disease control: lessons from ecology. Philos. Trans. R. Soc. Lond. B Biol. Sci. 372.

Colley, D.G., Andros, T.S., Campbell Jr., C.H., 2017. Schistosomiasis is more prevalent than previously thought: what does it mean for public health goals, policies, strategies, guidelines and intervention programs? Infect. Dis. Poverty 6, 63.

Colley, D.G., Binder, S., Campbell, C., King, C.H., Tchuem Tchuente, L.A., N'goran, E.K., Erko, B., Karanja, D.M., Kabatereine, N.B., Van Lieshout, L., Rathbun, S., 2013. A five-country evaluation of a point-of-care circulating cathodic antigen urine assay for the prevalence of *Schistosoma mansoni*. Am. J. Trop. Med. Hyg. 88, 426–432.

Cook, J.A., Jordan, P., Woodstock, L., Pilgrim, V., 1977. A controlled trial of hycanthone and placebo in schistosomiasis mansoni in St. Lucia. Ann. Trop. Med. Parasitol. 71, 197–202.

Crellen, T., Walker, M., Lamberton, P.H., Kabatereine, N.B., Tukahebwa, E.M., Cotton, J.A., Webster, J.P., 2016. Reduced efficacy of praziquantel against *Schistosoma mansoni* Is associated with multiple rounds of mass drug administration. Clin. Infect. Dis. 63, 1151–1159.

Diawara, A., Schwenkenbecher, J.M., Kaplan, R.M., Prichard, R.K., 2013. Molecular and biological diagnostic tests for monitoring benzimidazole resistance in human soil-transmitted helminths. Am. J. Trop. Med. Hyg. 88, 1052–1061.

Eisenbarth, A., Achukwi, M.D., Renz, A., 2016. Ongoing transmission of *Onchocerca volvulus* after 25 years of annual ivermectin mass treatments in the Vina du Nord River Valley, in north Cameroon. PLoS Negl. Trop. Dis. 10 e0004392.

Ezeamama, A.E., Bustinduy, A.L., Nkwata, A.K., Martinez, L., Pabalan, N., Boivin, M.J., King, C.H., 2018. Cognitive deficits and educational loss in children with schistosome infection-A systematic review and meta-analysis. PLoS Negl. Trop. Dis. 12 e0005524.

Farrell, S.H., Anderson, R.M., 2018. Helminth lifespan interacts with non-compliance in reducing the effectiveness of anthelmintic treatment. Parasites Vectors 11, 66.

Fenwick, A., Webster, J.P., Bosque-Oliva, E., Blair, L., Fleming, F.M., Zhang, Y., Garba, A., Stothard, J.R., Gabrielli, A.F., Clements, A.C., Kabatereine, N.B., Toure, S., Dembele, R., Nyandindi, U., Mwansa, J., Koukounari, A., 2009. The Schistosomiasis Control Initiative (SCI): rationale, development and implementation from 2002–2008. Parasitology 136, 1719–1730.

Fischer, P.U., King, C.L., Jacobson, J.A., Weil, G.J., 2017. Potential value of triple drug therapy with ivermectin, diethylcarbamazine, and albendazole (IDA) to accelerate elimination of lymphatic filariasis and onchocerciasis in Africa. PLoS Negl. Trop. Dis. 11 e0005163.

Garchitorena, A., Sokolow, S.H., Roche, B., Ngonghala, C.N., Jocque, M., Lund, A., Barry, M., Mordecai, E.A., Daily, G.C., Jones, J.H., Andrews, J.R., Bendavid, E., Luby, S.P., LaBeaud, A.D., Seetah, K., Guegan, J.F., Bonds, M.H., De Leo, G.A., 2017. Disease ecology, health and the environment: a framework to account for ecological and socio-economic drivers in the control of neglected tropical diseases. Philos. Trans. R. Soc. Lond. B Biol. Sci. 372.

Gardon, J., Gardon-Wendel, N., Demanga, N., Kamgno, J., Chippaux, J.P., Boussinesq, M., 1997. Serious reactions after mass treatment of onchocerciasis with ivermectin in an area endemic for *Loa loa* infection. Lancet 350, 18–22.

Gateff, C., Lemarinier, G., Labusquiere, R., Nebout, M., 1971. Influence de la bilharziose vesicale sur la rentabilite economique d'une population adulte jeune du Cameroun. Ann. Soc. Belge Med. Trop. 51, 309–324.

GBD 2016 Disease and Injury Incidence and Prevalence Collaborators, 2017. Global, regional, and national incidence, prevalence, and years lived with disability for 328 diseases and injuries for 195 countries, 1990-2016: a systematic analysis for the Global Burden of Disease Study 2016. Lancet 390, 1211–1259.

Gurarie, D., King, C.H., 2014. Population biology of *Schistosoma* mating, aggregation, and transmission breakpoints: more reliable model analysis for the end-game in communities at risk. PLoS One 9 e115875.

Haggag, A.A., Rabiee, A., Abd Elaziz, K.M., Gabrielli, A.F., Abdel Hay, R., Ramzy, R.M., 2017. Mapping of *Schistosoma mansoni* in the Nile Delta, Egypt: assessment of the prevalence by the circulating cathodic antigen urine assay. Acta Trop. 167, 9–17.

Hofstraat, K., Van Brakel, W.H., 2016. Social stigma towards neglected tropical diseases: a systematic review. Int. Health 8 (Suppl 1), i53–i70.

Hotez, P.J., Brindley, P.J., Bethony, J.M., King, C.H., Pearce, E.J., Jacobson, J., 2008. Helminth infections: the great neglected tropical diseases. J. Clin. Invest. 118, 1311–1321.

Ivy, J.A., King, C.H., Cook, J.A., Colley, D.G., 2018. Historical perspective: revisiting the St. Lucia Project, a multi-year comparison trial of schistosomiasis control strategies. PLoS Negl. Trop. Dis. 12 e0006223.

Jia, T.W., Melville, S., Utzinger, J., King, C.H., Zhou, X.N., 2012. Soil-transmitted helminth reinfection after drug treatment: a systematic review and meta-analysis. PLoS Negl. Trop. Dis. 6 e1621.

Kaisar, M.M.M., Brienen, E. a. T., Djuardi, Y., Sartono, E., Yazdanbakhsh, M., Verweij, J.J., Supali, T., van Lieshout, L., 2017. Improved diagnosis of *Trichuris trichiura* by using a bead-beating procedure on ethanol preserved stool samples prior to DNA isolation and the performance of multiplex real-time PCR for intestinal parasites. Parasitology 144, 965–974.

Kamgno, J., Pion, S.D., Chesnais, C.B., Bakalar, M.H., D'ambrosio, M.V., Mackenzie, C.D., Nana-Djeunga, H.C., Gounoue-Kamkumo, R., Njitchouang, G.R., Nwane, P., Tchatchueng-Mbouga, J.B., Wanji, S., Stolk, W.A., Fletcher, D.A., Klion, A.D., Nutman, T.B., Boussinesq, M., 2017. A Test-and-Not-Treat strategy for onchocerciasis in *Loa loa*-endemic areas. N. Engl. J. Med. 377, 2044–2052.

King, C.H., 2015. It's time to dispel the myth of "asymptomatic" schistosomiasis. PLoS Negl. Trop. Dis. 9 e0003504.

King, C.H., Dickman, K., Tisch, D.J., 2005. Reassessment of the cost of chronic helmintic infection: a meta-analysis of disability-related outcomes in endemic schistosomiasis. Lancet 365, 1561–1569.

King, C.H., Fairley, J.K., 2015. Tapeworms (cestodes). In: Bennett, J.E., Dolin, R., Blaser, M.J. (Eds.), Principles and Practice of Infectious Diseases, eighth ed. Elsevier Saunders, Philadelphia.

Krentel, A., Fischer, P.U., Weil, G.J., 2013. A review of factors that influence individual compliance with mass drug administration for elimination of lymphatic filariasis. PLoS Negl. Trop. Dis. 7, e2447.

Kvalsvig, J.D., 1986. The effects of schistosomiasis haematobium on the activity of school children. J. Trop. Med. Hyg. 89, 85–90.

Latham, M.C., Stephenson, L.S., Kurz, K.M., Kinoti, S.N., 1990. Metrifonate or praziquantel treatment improves physical fitness and appetite of Kenyan schoolboys with *Schistosoma haematobium* and hookworm infections. Am. J. Trop. Med. Hyg. 43, 170–179.

Leenstra, T., Coutinho, H.M., Acosta, L.P., Langdon, G.C., Su, L., Olveda, R.M., McGarvey, S.T., Kurtis, J.D., Friedman, J.F., 2006. *Schistosoma japonicum* reinfection after praziquantel treatment causes anemia associated with inflammation. Infect. Immun. 74, 6398–6407.

Leslie, J., Garba, A., Boubacar, K., Yaye, Y., Sebongou, H., Barkire, A., Fleming, F.M., Mounkaila, I., Adamou, S., Jackou, M.L., 2013. Neglected tropical diseases: comparison of the costs of integrated and vertical preventive chemotherapy treatment in Niger. Int. Health 5, 78–84.

Lo, N.C., Gurarie, D., Yoon, N., Coulibaly, J.T., Bendavid, E., Andrews, J.R., King, C.H., 2018. Impact and cost-effectiveness of snail control to achieve disease control targets for schistosomiasis. Proc. Natl. Acad. Sci. U. S. A. 115, E584–E591.

Lustigman, S., Prichard, R.K., Gazzinelli, A., Grant, W.N., Boatin, B.A., Mccarthy, J.S., Basanez, M.G., 2012. A research agenda for helminth diseases of humans: the problem of helminthiases. PLoS Negl. Trop. Dis. 6 e1582.

MacDonald, G., 1965. The dynamics of helminth infections, with special reference to schistosomes. Trans. R. Soc. Trop. Med. Hyg. 59, 489–506.

Maguire, J.H., 2015. Introduction to helminth infections. In: Bennett, J.E., Dolin, R., Blaser, M.J. (Eds.), Principles and Practice of Infectious Diseases, eighth ed. Elsevier Saunders, Philadelphia.

May, R.M., 1977. Togetherness among schistosomes: its effects on the dynamics of infection. Math. Biosci. 35, 301–343.

McCallum, H., Fenton, A., Hudson, P.J., Lee, B., Levick, B., Norman, R., Perkins, S.E., Viney, M., Wilson, A.J., Lello, J., 2017. Breaking beta: deconstructing the parasite transmission function. Philos. Trans. R. Soc. Lond. B Biol. Sci. 372.

McCreesh, N., Booth, M., 2013. Challenges in predicting the effects of climate change on *Schistosoma mansoni* and *Schistosoma haematobium* transmission potential. Trends Parasitol. 29, 548–555.

McGarvey, S.T., Aligui, G., Graham, K.K., Peters, P., Olds, G.R., Olveda, R., 1996. Schistosomiasis japonica and childhood nutritional status in northeastern Leyte, the Philippines: a randomized trial of praziquantel versus placebo. Am. J. Trop. Med. Hyg. 54, 498–502.

Miller-Fellows, S.C., Howard, L., Kramer, R., Hildebrand, V., Furin, J., Mutuku, F.M., Mukoko, D., Ivy, J.A., King, C.H., 2017. Cross-sectional interview study of fertility, pregnancy, and urogenital schistosomiasis in coastal Kenya: documented treatment in childhood is associated with reduced odds of subfertility among adult women. PLoS Negl. Trop. Dis. 11 e0006101.

Molyneux, D.H., Hotez, P.J., Fenwick, A., 2005. "Rapid-impact interventions": how a policy of integrated control for Africa's neglected tropical diseases could benefit the poor. PLoS Med. 2, e336.

Moser, W., Coulibaly, J.T., Ali, S.M., Ame, S.M., Amour, A.K., Yapi, R.B., Albonico, M., Puchkov, M., Huwyler, J., Hattendorf, J., Keiser, J., 2017. Efficacy and safety of tribendimidine, tribendimidine plus ivermectin, tribendimidine plus oxantel pamoate, and albendazole plus oxantel pamoate against hookworm and concomitant soil-transmitted helminth infections in Tanzania and Cote d'Ivoire: a randomised, controlled, single-blinded, non-inferiority trial. Lancet Infect. Dis. 17, 1162–1171.

Musuva, R., Shen, Y., Wei, X., Binder, S., Ivy, J.A., Secor, W.E., Montgomery, S.P., King, C.H., Mwinzi, P.N.M., 2017. Change in children's school behavior after mass administration of praziquantel for *Schistosoma mansoni* infection in endemic areas of western Kenya: a pilot study using the Behavioral Assessment System for Children (BASC-2). PLoS One 12 e0181975.

Nicholls, R.S., Duque, S., Olaya, L.A., Lopez, M.C., Sanchez, S.B., Morales, A.L., Palma, G.I., 2018. Elimination of onchocerciasis from Colombia: first proof of concept of river blindness elimination in the world. Parasites Vectors 11, 237.

Njenga, S.M., Mwandawiro, C.S., Wamae, C.N., Mukoko, D.A., Omar, A.A., Shimada, M., Bockarie, M.J., Molyneux, D.H., 2011. Sustained reduction in prevalence of lymphatic filariasis infection in spite of missed rounds of mass drug administration in an area under mosquito nets for malaria control. Parasites Vectors 4, 90.

Nokes, C., Mcgarvey, S.T., Shiue, L., Wu, G., Wu, H., Bundy, D.A., Olds, G.R., 1999. Evidence for an improvement in cognitive function following treatment of *Schistosoma japonicum* infection in Chinese primary schoolchildren. Am. J. Trop. Med. Hyg. 60, 556–565.

Olds, G., King, C., Hewlett, J., Olveda, R., Wu, G., Ouma, J., Peters, P., Mcgarvey, S., Odhiambo, O., Koech, D., 1999. Double-blind placebo-controlled study of concurrent administration of albendazole and praziquantel in schoolchildren with schistosomiasis and geohelminths. J. Infect. Dis. 179, 996–1003.

Ortu, G., Ndayishimiye, O., Clements, M., Kayugi, D., Campbell Jr., C.H., Lamine, M.S., Zivieri, A., Magalhaes, R.S., Binder, S., King, C.H., Fenwick, A., Colley, D.G., Jourdan, P.M., 2017. Countrywide reassessment of *Schistosoma mansoni* infection in Burundi using a urine-circulating cathodic antigen rapid test: informing the National Control Program. Am. J. Trop. Med. Hyg. 96, 664–673.

Pabalan, N., Singian, E., Tabangay, L., Jarjanazi, H., Boivin, M.J., Ezeamama, A.E., 2018. Soil-transmitted helminth infection, loss of education and cognitive impairment in school-aged children: a systematic review and meta-analysis. PLoS Negl. Trop. Dis. 12 e0005523.

Rao, R.U., Samarasekera, S.D., Nagodavithana, K.C., Dassanayaka, T.D.M., Punchihewa, M.W., Ranasinghe, U.S.B., Weil, G.J., 2017. Reassessment of areas with persistent lymphatic filariasis nine years after cessation of mass drug administration in Sri Lanka. PLoS Negl. Trop. Dis. 11 e0006066.

Savaya Alkalay, A., Rosen, O., Sokolow, S.H., Faye, Y.P., Faye, D.S., Aflalo, E.D., Jouanard, N., Zilberg, D., Huttinger, E., Sagi, A., 2014. The prawn *Macrobrachium vollenhovenii* in the Senegal River basin: towards sustainable restocking of all-male populations for biological control of schistosomiasis. PLoS Negl. Trop. Dis. 8 e3060.

Savioli, L., Daumiere, D., 2012. In: Crompton, D.W. (Ed.), Accelerating Work to Overcome the Global Impact of Neglected Tropical Diseases: A Roadmap for Implementation. World Health Organization, Geneva.

Sripa, B., Tangkawattana, S., Sangnikul, T., 2017. The Lawa model: a sustainable, integrated opisthorchiasis control program using the EcoHealth approach in the Lawa Lake region of Thailand. Parasitol. Int. 66, 346–354.

Stephenson, L.S., Kinoti, S.N., Latham, M.C., Kurz, K.M., Kyobe, J., 1989a. Single dose metrifonate or praziquantel treatment in Kenyan children. I. Effects on *Schistosoma haematobium*, hookworm, hemoglobin levels, splenomegaly, and hepatomegaly. Am. J. Trop. Med. Hyg. 41, 436–444.

Stephenson, L.S., Latham, M.C., Kurz, K.M., Kinoti, S.N., 1989b. Single dose metrifonate or praziquantel treatment in Kenyan children. II. Effects on growth in relation to *Schistosoma haematobium* and hookworm egg counts. Am. J. Trop. Med. Hyg. 41, 445–453.

Stephenson, L.S., Latham, M.C., Kurz, K.M., Kinoti, S.N., Oduori, M.L., Crompton, D.W., 1985. Relationships of *Schistosoma hematobium*, hookworm and malarial infections and metrifonate treatment to hemoglobin level in Kenyan school children. Am. J. Trop. Med. Hyg. 34, 519–528.

Taylor, A.B., Pica-Mattoccia, L., Polcaro, C.M., Donati, E., Cao, X., Basso, A., Guidi, A., Rugel, A.R., Holloway, S.P., Anderson, T.J., Hart, P.J., Cioli, D., Loverde, P.T., 2015. Structural and functional characterization of the enantiomers of the antischistosomal drug oxamniquine. PLoS Negl. Trop. Dis. 9 e0004132.

Theobald, S., Macpherson, E.E., Dean, L., Jacobson, J., Ducker, C., Gyapong, M., Hawkins, K., Elphick-Pooley, T., Mackenzie, C., Kelly-Hope, L.A., Fleming, F.M., Mbabazi, P.S., 2017. 20 years of gender mainstreaming in health: lessons and reflections for the neglected tropical diseases community. BMJ Glob. Health 2 e000512.

Toor, J., Turner, H.C., Truscott, J.E., Werkman, M., Phillips, A.E., Alsallaq, R., Medley, G.F., King, C.H., Anderson, R.M., 2018. The design of schistosomiasis monitoring and evaluation programmes: The importance of collecting adult data to inform treatment strategies for *Schistosoma mansoni*. PLoS Negl. Trop. Dis. (in press).

United Nations General Assembly, 2015. Resolution 70/1. Transforming Our World: The 2030 Agenda for Sustainable Development. United Nations, New York.

Vercruysse, J., Charlier, J., Van Dijk, J., Morgan, E.R., Geary, T., Von Samson-Himmelstjerna, G., Claerebout, E., 2018. Control of helminth ruminant infections by 2030. Parasitology 1–10.

Warren, K.S., 1982. Selective primary health care: strategies for control of disease in the developing world. I. Schistosomiasis. Rev. Infect. Dis. 4, 715–726.

WHO, 2003. Report of the WHO Informal Consultation on the Use of Praziquantel during Pregnancy and Albendazole/mebendazole in Children under 24 Months. WHO/CDC/CPE/PVC/2002.4. World Health Organization, Geneva.

WHO, 2006. Preventive Chemotherapy in Human Helminthiasis: Coordinated Use of Anthelminthic Drugs in Control Interventions: A Manual for Health Professionals and Programme Managers. World Health Organization, Geneva.

WHO, 2017a. Crossing the Billion. World Health Organization, Geneva.

WHO, 2017b. Guideline: Alternative Mass Drug Administration Regimens to Eliminate Lymphatic Filariasis. World Health Organization, Geneva.

WHO, 2018. Preventive Chemotherapy Databank [Online]. Available: http://www.who.int/neglected_diseases/preventive_chemotherapy/databank/en/index.html.

Wiegand, R.E., Mwinzi, P.N.M., Montgomery, S.P., Chan, Y.L., Andiego, K., Omedo, M., Muchiri, G., Ogutu, M.O., Rawago, F., Odiere, M.R., Karanja, D.M.S., Secor, W.E., 2017. A persistent hotspot of *Schistosoma mansoni* infection in a five-year randomized trial of praziquantel preventative chemotherapy strategies. J. Infect. Dis. 216, 1425–1433.

Woolhouse, M.E., Etard, J.F., Dietz, K., Ndhlovu, P.D., Chandiwana, S.K., 1998. Heterogeneities in schistosome transmission dynamics and control. Parasitology 117, 475–482.

Zhou, X.N., Yang, G.J., Yang, K., Wang, X.H., Hong, Q.B., Sun, L.P., Malone, J.B., Kristensen, T.K., Bergquist, N.R., Utzinger, J., 2008. Potential impact of climate change on schistosomiasis transmission in China. Am. J. Trop. Med. Hyg. 78, 188–194.

CHAPTER THREE

Is *Opisthorchis viverrini* Emerging in Cambodia?

Virak Khieu*[,a], Thomas Fürst[§,¶,||,a], Kazuko Miyamoto[#], Tai-Soon Yong, Jong-Yil Chai[§§,¶¶], Rekol Huy*, Sinuon Muth* and Peter Odermatt[§,¶,1]**

*National Center for Parasitology, Entomology and Malaria Control, Phnom Penh, Cambodia
[§]Swiss Tropical and Public Health Institute, Basel, Switzerland
[¶]University of Basel, Basel, Switzerland
[||]School of Public Health, Imperial College London, London, United Kingdom
[#]School of Nursing, Faculty of Medicine & Center for International Education and Exchange, Yamanashi University, Yamanashi, Japan
**Department of Environmental Biology, Institute of Tropical Medicine, and Arthropods of Medical Importance Resource Bank, Yonsei University College of Medicine, Seoul, Republic of Korea
[§§]Institute of Parasitic Diseases, Korea Association of Health Promotion, Seoul, Republic of Korea
[¶¶]Department of Parasitology and Tropical Medicine, Seoul National University College of Medicine, Seoul, Republic of Korea
[1]Corresponding author: E-mail: peter.odermatt@swisstph.ch

Contents

[a] These authors contributed equally to this manuscript.

Advances in Parasitology, Volume 103
ISSN 0065-308X
https://doi.org/10.1016/bs.apar.2019.02.002

Abstract

Opisthorchis viverrini infection is widely prevalent in Southeast Asia. In Cambodia information on this helminth infection is scare. Recent reports suggest that *O. viverrini* is an emerging public health problem. We aimed to synthesize all information in relation to the infection, epidemiology, and morbidity of *O. viverrini* in Cambodia; from published as well as thus far unpublished sources. First reports on *O. viverrini* date back to 1995. In 2006 an *O. viverrini* initiative was launched by the national helminth control program. Since then *O. viverrini* has been reported in all — except two — provinces. Villages with high prevalences (>20%) were found in provinces from Preah Vihear to Takeo. The infection has a highly focal distribution. In many villages no infections were detected. *O. viverrini* infection was also reported in cats, dogs and intermediate hosts. No report on morbidity associated with *O. viverrini* was found. The current evidence suggests that *O. viverrini* infection remains underreported in Cambodia. It is likely that the transmission will further increase in the future with potentially serious consequences for human health.

1. INTRODUCTION

The small Asian liver fluke *Opisthorchis viverrini* is endemic in Southeast Asia (Sithithaworn et al., 2012a). While in Thailand and Laos many studies of recent years documented high infection rates and also a substantial burden in humans and animal intermediate and reservoir hosts, reports on *O. viverrini* and small flukes exist much less from Cambodia (Fürst et al., 2012a, 2012b). As the existing evidence on the epidemiology of *O. viverrini* in Cambodia has not yet been consistently synthesised, the public health relevance of this helminth infection for the country is not well understood.

The parasitic trematode *O. viverrini* has a complex life cycle. In addition to the mammalian definitive hosts of humans, cats and dogs, it involves two aquatic intermediate hosts, namely freshwater snails of the genus *Bithynia* and freshwater fish of the *Cyprinidae* family. Humans and other mammals are infected by eating raw or undercooked contaminated fish (Kaewkes, 2003). The hermaphrodite adult parasites dwell in the bile ducts where they develop and produce eggs. These eggs are transported with the bile to the intestine and released with the faeces. Poor sanitary conditions favour the contamination of water bodies with parasite eggs. In the water, miracidiae hatch from the eggs and are ingested by intermediate snails. After a period of development within the snails, infectious cercariae are released. These cercariae infect *Cyprinoid* fish, encyst as metacercariae under the skin, and infect again mammalian definitive hosts if they eat the fish containing viable parasite metacercariae. Hence, *O. viverrini* is transmitted in communities where the consumption of raw or insufficiently cooked fish dishes is deeply culturally rooted (Xayaseng et al., 2013). In infected humans, *O. viverrini* may provoke severe hepatobiliary morbidity (Mairiang and Mairiang, 2003) including cholangiocarcinoma (CCA), an aggressive and fatal bile duct cancer (Sripa et al., 2012).

Raw and insufficiently cooked fish consumption is a habit observed in many communities of rural Cambodia. However, it is not clear whether these eating habits are sufficiently consistent and whether infection with suitable intermediate molluscs and fish hosts are sufficiently present so transmission foci may widely develop. Furthermore, it takes a considerable transmission period and resulting infection intensity for the development of human hepatobiliary morbidity.

In this contribution, we aim to bring together all information in relation to the infection, epidemiology, and morbidity of *O. viverrini* in Cambodia from published as well as thus far unpublished sources. With this work, we want to compile the current evidence base, provide guidance to the national helminth control programme in Cambodia, and – if need be – contribute to the development of a comprehensive Cambodian intervention strategy against potentially severe *O. viverrini* infections.

2. METHODS

The review and compilation work consisted of two parts. First, five online literature databases were systematically searched for peer-reviewed and published articles and the results cross-checked and complemented by references from an already existing database on food-borne trematodes, which the authors had previously compiled for other projects (for further details see the following Section 2.1). Second, unpublished reports from the National Centre for Parasitology, Entomology and Malaria Control (CNM) and personal archives from our operational research activities in Cambodia from 1995 to 2015 were screened for relevant information on *O. viverrini* and other small liver fluke infections (for further details see the following Section 2.2). The obtained information was then categorised and evaluated as further detailed in Section 2.3 below.

2.1 Systematic Review of the Published Literature

A broad-based, computer-aided systematic review of the published literature was conducted with the aim to identify all available relevant information about opisthorchiasis in Cambodia. However, still today, considerable difficulties prevail in unambiguously diagnosing *O. viverrini* infections. The most common approach is direct parasitological diagnosis by parasite egg detection in faecal samples (e.g. Kato-Katz [K-K] thick smear, Stoll's dilution egg count method, formalin-ethyl-acetate technique, or sedimentation techniques), but *O. viverrini* eggs are difficult to differentiate from eggs of *Clonorchis sinensis*, another food-borne liver fluke, and eggs of small intestinal food-borne trematodes (e.g. species of the families Gymnophallidae, Heterophyidae, Lecithodendriidae, or Plagiorchiidae; see also Appendix 1, Table A1, for an overview of the taxonomy of food-borne trematodes) (Johansen et al., 2010; Fürst et al., 2012c; Wongsawad and Wongsawad, 2012; Clausen and Murrell, 2014; Maguire, 2015). Because of these diagnostic challenges, studies sometimes simply report "*Opisthorchis*-like eggs", "small fluke infections", or "minute fluke infections". To adjust for this ambiguity, we decided to adapt a search strategy, which targeted all food-borne trematode species, and to combine the respective search terms with "Cambodia".

Five bibliographic databases (PubMed, ISI Web of Science, Science Direct, WHOLIS, and FAOBIB) were searched online using Boolean operators, wildcards, respective thesauri, whenever available, and normal keyword searches including specific parasite and disease names otherwise (again, for an overview over the taxonomy of food-borne trematodes see Appendix 1, Table A1). No language and no time restrictions were set. The search was performed at the end of 2015. The details of this search strategy are summarized in Table 1.

In order to further enhance our evidence base, an existing database, which resulted from one of the authors' involvement in the Global Burden of Disease (GBD) Collaboration and contained more than 41,000 raw hits pertaining to food-borne trematodiasis (Fürst et al., 2012a, 2012b, 2012c), was screened for manuscripts with reference to Cambodia. Furthermore, the bibliographies of all potentially relevant articles were hand-searched for additional references.

2.2 Evidence From Unpublished Reports and Personal Archives of the Authors

Unpublished reports of the last 20 years from CNM (1995–2014) and personal archives from our operational helminthiasis research activities in Cambodia were searched for information on *O. viverrini* and small liver fluke infection. As a first source of additional relevant data, the results of the

Table 1 Summary of the Search Strategy and Selection Criteria of the Systematic Review. The table highlights the databases searched, respective thesauri, search terms, and Boolean operators used, and numbers of resulting hits. The numbers of resulting hits represents the literature as per the end of 2015.

Database	Thesaurus (provider)	Search term (OR)	Inclusion (AND)	Number of hits
PubMed	MeSH (US National Library of Medicine)	"trematode infections" "trematoda"	Cambodia	59
ISI Web of Science	no thesaurus: normal keyword search	"trematod[b]", "food-borne", "plant-borne", "fish-borne", "fluke[b]", specific species and disease names[a]	Cambodia	133
Science Direct	no thesaurus: normal keyword search	"trematod[b]", "food-borne", "plant-borne", "fish-borne", "fluke[b]", specific species and disease names[a]	Cambodia	525
WHO (WHOLIS)	MeSH (US National Library of Medicine)	"trematode infections" "trematoda"	—	17
FAO (FAOBIB)	AGROVOC (FAO)	"helminthoses" "trematoda"	—	142
Total				876

[a]See Appendix 1, Table A1.
[b]Wildcard; represents any group of characters, including no character.

annual *Schistosoma mekongi* surveillance were identified. This schistosomiasis surveillance was conducted between 1995 and 2010 in the general population of selected villages in Sambo and Kratie districts, Kratie province. The surveillance used the K-K diagnostic technique (Katz et al., 1972) to identify *S. mekongi* eggs in a single stool sample and, as a by-product, *O. viverrini* infections were identified. Also of note, after the surveillance activities, an annual praziquantel mass-drug administration was conducted in all identified endemic villages in the Kratie province.

Also as a part of the *S. mekongi* surveillance, three cross-sectional surveys were conducted among schoolchildren in 2000 in Kratie province and the further downstream and upstream provinces along the Mekong river, Kampong Cham and Stung Treng. The formalin-detergent sedimentation technique was applied for diagnosis. Again, human *O. viverrini* infections were recorded as a by-product.

In 2005, three cross-sectional surveys were conducted in Takeo province in pregnant women, schoolchildren, and school teachers. These surveys were conducted in 8—9 villages, used the K-K diagnostic technique (Katz et al., 1972), and also produced thus far unpublished data on human *O. viverrini* infections.

A community-based study on intestinal parasite infections was identified as a second source of additional relevant data. The study was carried out amongst the general population living in 45 villages of 5 provinces (Kampot, Modulkiri, Ratanakiri, Stung Treng, Takeo) in southern and north eastern Cambodia in 2007. Again, the K-K diagnostic technique (Katz et al., 1972) was employed and *O. viverrini* eggs were recorded also.

In 2010, a small cross-sectional survey was conducted among adults in one village in Prey Veng province. The unpublished data on *O. viverrini* originates from the diagnosis with the K-K technique (Katz et al., 1972).

Another source of additional relevant data stems from a survey in the general population of 60 villages in Preah Vihear province. This survey was conducted in 2010, used the formalin-ether concentration technique, and also diagnosed human *O. viverrini* infections in a subsample of 20 villages.

Also in 2010, a small survey specifically aimed to explore the presence of *O. viverrini* infections in a suspected high-risk village was conducted in Kampong Samy in the Chhaeb district, Preah Vihear province located along the Mekong River at the border to Laos where the habit of eating undercooked/raw fish was reported to be very common. The survey included 70 villagers from the general population older than 2 years.

As another source of data, a study from 2011 was identified, assessed the knowledge, attitudes and practices regarding intestinal helminth infections in the general population of 4 highly schistosomiasis endemic villages of Kratie province. Unlike most other surveys, this one diagnosed human intestinal helminthiasis not only with the K-K technique (Katz et al., 1972) on a single stool sample, but also with the formalin-ether concentration technique.

Three cross-sectional surveys in adults in Kampong Cham, Kandal, and Ratanakiri provinces were completed in 2012. These surveys mainly relied on the K-K technique (Katz et al., 1972) and generated data on human opisthorchiasis.

A nation-wide survey on intestinal parasites of schoolchildren was conducted in 2013 after providing regular, twice annual mebendazole and/or albendazole mass drug administration (MDA) in schools and communities. For the survey, one primary school of each province was randomly selected from the school list. Around 70 schoolchildren per primary school were requested to submit one stool sample, which was then examined with the K-K technique (Katz et al., 1972) and — again amongst other parasites — *O. viverrini* diagnosed.

Finally, a set of cross-sectional surveys in adults was completed in 2013 in the three provinces of Tbaung Kmom, Kampong Thom, and Kampong Cham. The thus far unpublished data on human *O. viverrini* diagnosis was obtained by applying not only the K-K technique but also polymerase chain reaction (PCR).

2.3 Categorisation and Evaluation of the Obtained Information

All quantitative primary data on human or animal opisthorchiasis in Cambodia from published or unpublished sources were retrieved and compiled in tables. As meta-information —to better understand and correctly interpret the primary quantitative data — on study duration, study locations, study types, study populations, and applied diagnostic techniques were recorded. These data were then used to produce overview maps; one on the human and one on the animal evidence of *O. viverrini* epidemiology in Cambodia. The identified non-quantitative (i.e. qualitative) information was compared with the quantitative data and used to further verify our understanding, interpretation, and evaluation of the situation in the Discussion section.

3. RESULTS

3.1 Evidence From the Published Literature

Our initial systematic review resulted in 876 raw hits (Table 1, Fig. 1). From the screening of the existing GBD database, an additional 74 raw hits were retrieved. Together, and after exclusion of all duplicates and all references considered not relevant based on title and abstract, 247 potentially relevant references remained. An additional 77 potentially relevant references were identified by hand-searching the bibliographies of already identified articles. Eleven of the thus obtained 324 potentially relevant references were not accessible. The full-text review of all remaining references revealed that 295 contained no relevant quantitative primary data regarding the prevalence of *O. viverrini* or other small fluke infections in Cambodia. Nevertheless, the relevant qualitative information from these 295 references was considered in the Discussion. The remaining 18 references contained relevant quantitative primary data on human (total 13 studies summarized

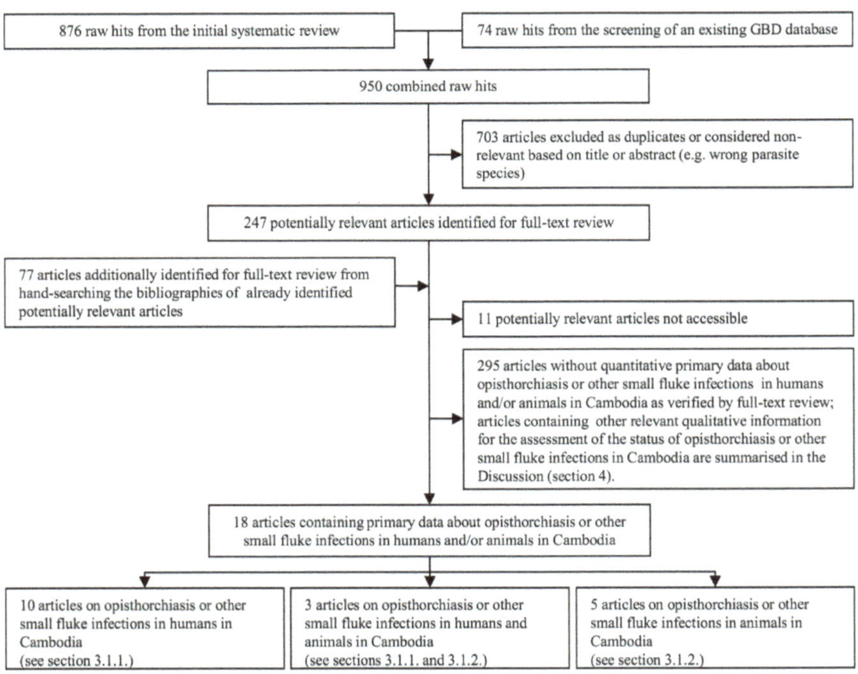

Figure 1 ***Flowchart highlighting the outcome of the systematic review of the published literature.*** For a summary of the applied search strategy and selection criteria see Table 1.

in Section 3.1.1.) and/or animal (total 8 studies summarized in Section 3.1.2.) infection with *O. viverrini*/small flukes in Cambodia. All the relevant quantitative primary data were on prevalences only.

3.1.1 Published Studies on Opisthorchiasis in Humans in Cambodia

Thirteen studies reported opisthorchiasis or other small fluke infections in humans in all except three provinces (Kampong Speu, Kep, Prey Veng) of Cambodia (Figs. 1 and 2, Table 2) (Kobayashi et al., 1999; Lee et al., 2002; Muth et al., 2003; Sohn et al., 2011a, 2011b, 2012; Yong et al., 2012, 2014; Khieu et al., 2013, 2014; Miyamoto et al., 2014; Schär et al., 2014; Bless et al., 2015). According to these studies, a geographic strip with higher prevalences of more than 5% seems to run through the centre of Cambodia from North (Preah Vihear province) to South (Takeo

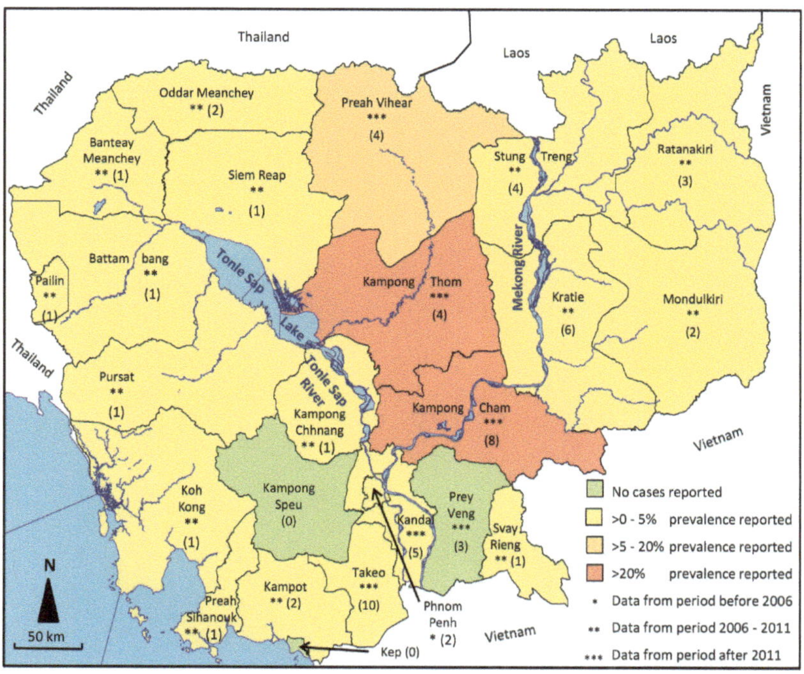

Figure 2 ***Map highlighting the prevalences of opisthorchiasis/small fluke infections in humans in Cambodia as reported in the published and unpublished literature.*** In case of more than one report on the prevalence in the same district, firstly, surveys in the general population, secondly, more recent surveys, and, thirdly, surveys with larger study samples were given priority for illustration in the map. The numbers in brackets indicate the total number of surveys identified in each district. Details about the data sources are provided in Tables 2 and 4.

Table 2 Identified Primary Data From the Published Literature on Human Opisthorchiasis/Small Fluke Infections in Cambodia. The studies are listed in chronological order (year of start of study) for each province.

Location (province)	References	Parasite species reported	Year of study	Study type	Study population	No. positive/no. examined (prevalence)	Diagnostic Technique	Remarks
Banteay Meanchey Province	Yong et al. (2014)	"Small trematode eggs"	2006–11	Cross-sectional	General population	5/820 (0.6%)	K-K	
Battambang Province	Yong et al. (2014)	"Small trematode eggs"	2006–11	Cross-sectional	General population	2/807 (0.2%)	K-K	
Kampong Cham Province	Lee et al. (2002)	*Opisthorchis viverrini*	2002	Cross-sectional	Schoolchildren	10/251 (4.0%)	FEC	
	Yong et al. (2014)	"Small trematode eggs"	2006–11	Cross-sectional	General population	622/2,593 (24.0%)	K-K	
	Miyamoto et al. (2014)	*Opisthorchis viverrini*	2010	Cross-sectional	Adults	108/379 (28.5%)	K-K	All households
	Miyamoto et al. (2014)	*Opisthorchis viverrini*	2010	Cross-sectional	Adults	322/1358 (23.7%)	K-K	Quasi-random
	Miyamoto et al. (2014)	*Opisthorchis viverrini*	2012	Cross-sectional	Adults	157/658 (23.9%)	K-K, PCR, PUR	

Kampong Chhnang Province	Yong et al. (2014)	"Small trematode eggs"	2006–11	Cross-sectional	General population	10/1,795 (0.6%)	K-K	
Kampong Thom Province	Yong et al. (2014)	"Small trematode eggs"	2006–11	Cross-sectional	General population	11/846 (1.3%)	K-K	
	Miyamoto et al. (2014)	*Opisthorchis viverrini*	2012	Cross-sectional	Adults	100/287 (34.8%)	K-K, PCR, PUR	
Kampot Province	Yong et al. (2014)	"Small trematode eggs"	2006–11	Cross-sectional	General population	1/2,732 (0.1%)	K-K	
Kandal Province	Miyamoto et al. (2014)	*Opisthorchis viverrini*	2008	Cross-sectional	Adults	161/797 (20.2%)	K-K, PCR, PUR	All households
	Miyamoto et al. (2014)	*Opisthorchis viverrini*	2008	Cross-sectional	Adults	11/615 (1.8%)	K-K	All households
	Khieu et al. (2013)	"Small trematode eggs"	2009	Cross-sectional	Schoolchildren	36/458 (7.9%)	K-K	
	Bless et al. (2015)	*Opisthorchis viverrini*	2011	Cross-sectional	Schoolchildren	13/228 (5.7%)	K-K, FEC	Mean EPG: 38.7
Koh Kong Province	Yong et al. (2014)	"Small trematode eggs"	2006–11	Cross-sectional	General population	16/2,902 (0.6%)	K-K	

(*Continued*)

Table 2 Identified Primary Data From the Published Literature on Human Opisthorchiasis/Small Fluke Infections in Cambodia. The studies are listed in chronological order (year of start of study) for each province.—cont'd

Location (province)	References	Parasite species reported	Year of study	Study type	Study population	No. positive/no. examined (prevalence)	Diagnostic Technique	Remarks
Kratie Province	Yong et al. (2014)	"Small trematode eggs"	2006–11	Cross-sectional	General population	103/3,034 (3.4%)	K-K	
	Sohn et al. (2012)	*Opisthorchis viverrini*	2011	Cross-sectional	General population	96/2,101 (4.6%)	K-K, PCR, PUR	
Mondulkiri Province	Yong et al. (2014)	"Small trematode eggs"	2006–11	Cross-sectional	General population	5/837 (0.6%)	K-K	
Oddar Meanchey Province	Yong et al. (2014)	"Small trematode eggs"	2006–11	Cross-sectional	General population	32/2,449 (1.3%)	K-K	
	Sohn et al. (2011a)	"Small trematode eggs"	2007–09	Cross-sectional	General population	9/1,287 (0.7%)	K-K	
Pailin Province	Yong et al. (2014)	"Small trematode eggs"	2006–11	Cross-sectional	General population	1/793 (0.1%)	K-K	
Phnom Penh Municipality	Muth et al. (2003)	*Opisthorchis viverrini*	1996–99	Cross-sectional	Schoolchildren	2/1,535 (0.1%)	K-K	
	Kobayashi et al. (1999)	*Opisthorchis viverrini*	1998	Cross-sectional	General population	4/113 (3.5%)	K-K, FEC	

Preah Sihanouk Province	Yong et al. (2014)	"Small trematode eggs"	2006–11	Cross-sectional	General population	22/3,807 (0.6%)	K-K	
Preah Vihear Province	Yong et al. (2014)	"Small trematode eggs"	2006–11	Cross-sectional	General population	23/862 (2.7%)	K-K	
	Schär et al. (2014)	"Small trematode eggs"	2012	Cross-sectional	General population	11/218 (5.1%)	K-K, FEC, FLO	Mean EPG: 36.0
Prey Veng	Miyamoto et al. (2014)	*Opisthorchis viverrini*	2010–12	Cross-sectional	Adults	0/759 (0.0%)	K-K	
Pursat Province	Yong et al. (2014)	"Small trematode eggs"	2006–11	Cross-sectional	General population	3/770 (0.4%)	K-K	
Ratanakiri Province	Yong et al. (2014)	"Small trematode eggs"	2006–11	Cross-sectional	General population	27/959 (2.8%)	K-K	
Siem Reap Province	Yong et al. (2014)	"Small trematode eggs"	2006–11	Cross-sectional	General population	2/493 (0.4%)	K-K	
Stung Treng Province	Muth et al. (2003)	*Opisthorchis viverrini*	1998	Cross-sectional	Schoolchildren	8/341 (2.3%)	K-K	
	Yong et al. (2014)	"Small trematode eggs"	2006–11	Cross-sectional	General population	24/975 (2.5%)	K-K	

(Continued)

Table 2 Identified Primary Data From the Published Literature on Human Opisthorchiasis/Small Fluke Infections in Cambodia. The studies are listed in chronological order (year of start of study) for each province.—cont'd

Location (province)	References	Parasite species reported	Year of study	Study type	Study population	No. positive/no. examined (prevalence)	Diagnostic Technique	Remarks
Svay Rieng Province	Yong et al. (2014)	"Small trematode eggs"	2006–11	Cross-sectional	General population	5/838 (0.6%)	K-K	
Takeo Province	Miyamoto et al. (2014)	*Opisthorchis viverrini*	2006	Cross-sectional	Adults	352/807 (43.6%)	K-K	All households
	Miyamoto et al. (2014)	*Opisthorchis viverrini*	2006–07	Cross-sectional	Adults	21/10,422 (0.2%)	K-K	All households
	Yong et al. (2012)	*Opisthorchis viverrini*	2006–10	Cross-sectional	General population	855/1,799 (47.5%)	K-K	
	Yong et al. (2014)	"Small trematode eggs"	2006–11	Cross-sectional	General population	924/3,889 (23.8%)	K-K	
	Sohn et al. (2011b)	"Small trematode eggs"	2010	Cross-sectional	General population	646/1,993 (32.4%)	–	
	Khieu et al. (2014)	"Small trematode eggs"	2011	Cross-sectional	General population	6/2,861 (0.2%)	K-K	

–, no details provided; BAER, Baermann technique; ELISA, Enzyme-linked immunosorbent assay; EPG, Eggs per gram of stool; FEC, Formalin-ether concentration technique; FLO, Flotation method; KAP, Koga Agar plate culture; K-K, Kato-Katz technique; PCR, Polymerase chain reaction; PUR, Purging.

province) with lower prevalence provinces in the East and West of the country (Fig. 2). Interestingly, however, the northern part of the high prevalence strip lies in the West of the Mekong River and does not include the provinces Stung Treng and Kratie along the Mekong River. In the centre, Kampong Chhnang seems to be a low prevalence province despite its location along the Tonle Sap River. In the South, the low prevalences reported for the Phnom Penh Municipal Province and the absence of any cases in Kampong Speu and Prey Veng, despite their location at or in the vicinity of the Mekong River, is remarkable.

As summarised in Table 2, all identified studies were conducted in the last 20 years, had a cross-sectional study design, mostly focused on the general population and used K-K as diagnostic technique. The study samples ranged from 113 individuals in a study in Phnom Penh (Kobayashi et al., 1999) to 11,229 in a study in Takeo (Miyamoto et al., 2014) and identified prevalences varied between 0 and 0.1% (Muth et al., 2003; Yong et al., 2014) and 47.5% (Yong et al., 2012). Only two studies reported infection intensities (mean numbers of parasite eggs per gram stool examined [EPG]). Across all infected individuals, these two studies found with the K-K technique EPGs of 38.7 in Kandal (Bless et al., 2015) and of 36.0 in Preah Vihear (Schär et al., 2014), which are low infection intensities according to Maleevong and colleagues (Maleewong et al., 1992).

Overall, the studies document a highly focal distribution of *O. viverrini* infection; e.g. in 2006 in two villages of Takeo province 43.6% of 807 examined villagers were positive. In the following 16 villages investigated only 21 cases were found in 10′422 individuals (0.2%); all these cases occurred in 2 of the 16 villages (Miyamoto et al., 2014).

3.1.2 Published Studies on Opisthorchiasis in Animals in Cambodia

Eight manuscripts containing primary data about opisthorchiasis or other small fluke infections in animals in Cambodia were identified (Figs. 1 and 3, Table 3) (Touch et al., 2009, 2013; Ngoenkla et al., 2010; Thaenkham et al., 2010; Sohn et al., 2011b, 2012; Chai et al., 2014; Miyamoto et al., 2014). One study investigated infections in first (snails) and second (freshwater fish) intermediate hosts (Miyamoto et al., 2014) and seven focused exclusively on infections in second (various species of freshwater fish) intermediate hosts (Touch et al., 2009, 2013; Ngoenkla et al., 2010; Thaenkham et al., 2010; Sohn et al., 2011b, 2012; Chai et al., 2014) (Table 3). All studies were completed between 2006 and 2012, had a cross-sectional study design and used varying diagnostic techniques.

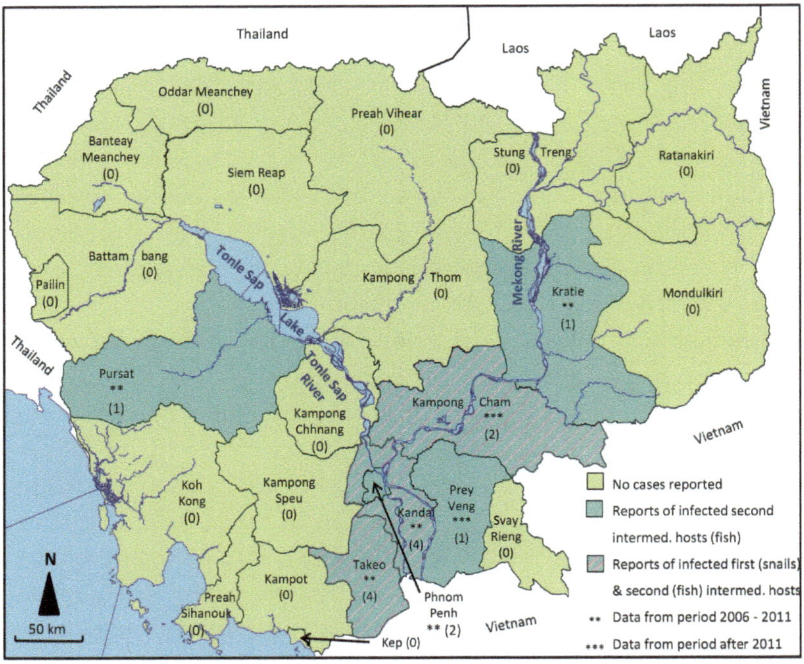

Figure 3 ***Map highlighting the prevalences of opisthorchiasis/small fluke infections in animals in Cambodia as reported in the published and unpublished literature.*** In case of more than one report on the prevalence in the same district, firstly, more recent surveys, and, secondly, surveys with larger study samples were given priority for illustration in the map. The numbers in brackets indicate the total number of surveys identified in each district. Details about the data sources are provided in Table 3.

Reported prevalences were 0.2% in snails (Miyamoto et al., 2014) and 5.7–33.3% in fish (Touch et al., 2009, 2013; Sohn et al., 2011b, 2012; Chai et al., 2014). The average number of cercariae identified per infected snail was 1.0 (Miyamoto et al., 2014) and reports on the average number of metacercariae per infected fish ranged between 1.5 and 19.3 in the studies (Touch et al., 2009; Sohn et al., 2011b, 2012; Chai et al., 2014). Furthermore, two studies noted that metacercarial burden in fish was higher in the dry season than in the wet season (Ngoenkla et al., 2010; Touch et al., 2013) and one study also calculated the average number of metacercariae per kilogramme of infected fish as 125.6 (Ngoenkla et al., 2010).

All reports of animal infections are concentrated in the South and South-West of Cambodia with one notable exception (Fig. 3, Table 3) (Touch et al., 2009, 2013; Ngoenkla et al., 2010; Thaenkham et al., 2010; Sohn et al., 2011b, 2012; Chai et al., 2014; Miyamoto et al., 2014). *O. viverrini*

Table 3 Identified Primary Data From the Published Literature on Animal Opisthorchiasis/Small Fluke Infections in Cambodia. The studies are listed in chronological order (year of start of study) for each province.

Location (province)	References	Parasite species reported	Year of study	Study type	Study population	No. positive/ no. examined (prevalence)	Diagnostic technique	Remarks
Kampong Cham Province	Miyamoto et al. (2014)	*Opisthorchis viverrini*	2010–12	Cross-sectional	6 species of freshwater fish	–	Morphology, animal experiment, PCR	
	Miyamoto et al. (2014)	*Opisthorchis viverrini*	2010–12	Cross-sectional	*Bithynia* spp. snails	1/406 (0.2%)	Morphology, PCR	average no. cercariae/ infected snail 1.0
Kandal and Takeo Province	Touch et al. (2009)	*Opisthorchis viverrini*	2007–08	Cross-sectional	30 species of freshwater fish	163/1,479 (11.0%)	Compression method and and animal experiment	average no. metacercariae/ infected fish 4.8
Kandal Province	Miyamoto et al. (2014)	*Opisthorchis viverrini*	2008	Cross-sectional	*Bithynia* spp. snails	–	Morphology, PCR	
	Touch et al. (2013)	*Opisthorchis viverrini*	2008–09	Cross-sectional	17 species of freshwater fish	624/1,874 (33.3%)	Artificial digestion and compression method	metacercarial burden higher in dry season

(*Continued*)

Table 3 Identified Primary Data From the Published Literature on Animal Opisthorchiasis/Small Fluke Infections in Cambodia. The studies are listed in chronological order (year of start of study) for each province.—cont'd

Location (province)	References	Parasite species reported	Year of study	Study type	Study population	No. positive/no. examined (prevalence)	Diagnostic technique	Remarks
	Thaenkham et al. (2010)	*Opisthorchis viverrini*	2010[a]	Cross-sectional	1 species of freshwater fish	—	Artificial digestion and animal experiment	
Kratie Province	Sohn et al. (2012)	*Opisthorchis viverrini*	2011	Cross-sectional	7 species of freshwater fish	19/85 (22.4%)	Artificial digestion and animal experiment	average no. metacercariae/infected fish 19.3
Takeo Province	Miyamoto et al. (2014)	*Opisthorchis viverrini*	2006–07	Cross-sectional	*Bithynia* spp. snails	—	Morphology, PCR	
	Miyamoto et al. (2014)	*Opisthorchis viverrini*	2006–07	Cross-sectional	6 species of freshwater fish	—	Morphology, animal experiment, PCR	
	Sohn et al. (2011b)	*Opisthorchis viverrini*	2010	Cross-sectional	2 species of freshwater fish	5/15 (33.3%)	Artificial digestion and animal experiment	average no. metacercariae/infected fish 10.0

Phnom Penh Municipality	Chai et al. (2014)	*Opisthorchis viverrini*	2006–10	Cross-sectional	22 species of freshwater fish	37/184 (20.1%)	Artificial digestion and sedimentation	Average no. metacercariae/infected fish 18.6
	Ngoenkla et al. (2010)	*Opisthorchis viverrini*	2007–08	Cross-sectional	15 species of freshwater fish	–	Artificial digestion and sedimentation	average no. metacercariae/kg infected fish 125.6; metacercarial
Prey Veng Province	Miyamoto et al. (2014)	*Opisthorchis viverrini*	2010–12	Cross-sectional	6 species of freshwater fish	–	Morphology, animal experiment, PCR	
Pursat Province	Chai et al. (2014)	*Opisthorchis viverrini*	2007	Cross-sectional	5 species of freshwater fish	2/36 (5.6%)	Artificial digestion and sedimentation	average no. metacercariae/infected fish 1.5

–, no details provided; PCR, Ppolymerase chain reaction.
[a]No year of study mentioned and hence, the year of publication is provided.

and four other small flukes (*Haplorchis yokogawai*, *Haplorchis pumilio*, *Centrocestus formosanus*, *Procerovum* sp.) were reported from fish in Pursat province in the West of the country and bordering also Tonle Sap Lake (Chai et al., 2014). All other reports came from provinces in the vicinity of the Mekong River, with both snail and fish infections identified in Kampong Cham, Kandal and Takeo, and only fish infections mentioned in Kratie, Prey Veng, and Phnom Penh. Interestingly, no animal case reports could be obtained from Stung Treng along the Mekong River, in the North of Cambodia bordering Laos, and from Kampong Chhnang along the Tonle Sap River.

3.2 Evidence From Unpublished Reports

The quantitative primary data on opisthorchiasis in Cambodia from unpublished reports of the last two decades from the CNM (1995–2014) and personal archives is summarized in Table 4. As mentioned in Section 2.2, several additional and unpublished data sources were unravelled. All the unpublished data focused on humans and provided prevalence data only. As indicated also in the respective legends, the maps in Figs. 2 and 3 sequentially prioritised (i) surveys in the general population, (ii) more recent surveys, and (iii) surveys with larger study samples, if more than one report on the prevalence in the same district could be identified. Based on this procedure, only two changes emerged in the maps due to the complementation of the published data. First, the human prevalence in Takeo province was revised from 23.8% (Table 2) according to Yong and colleagues (Yong et al., 2014) to 0.6% (yellow colouring in Fig. 2; see also Table 4). Second, the human prevalence in Kandal province was reduced from 5.7% (Table 2) according to Bless and colleagues (Bless et al., 2015) to 4.2% (yellow colouring in Fig. 2; see also Table 4).

With the exception of one source, which was from sentinel site surveillance, all unpublished studies were simple cross-sectional studies. Fourteen of the additional and unpublished data points considered the general population as main study population, 7 focused on adults and 5 on schoolchildren, and one study each examined pregnant women and school teachers (Table 4). However, the respective study samples largely varied from 58 to 2,500 individuals and, unsurprisingly, the *O. viverrini* infection rates ranged from zero to 58.4%.

Interestingly, in a first cross-sectional survey in 1996–97, the prevalence of *O. viverrini* infection in the general population in selected *S. mekongi* endemic villages of Kratie was examined and found to be high (18%).

Table 4 Identified Unpublished Primary Data on Human Opisthorchiasis/Small Fluke Infections in Cambodia. The studies are listed in chronological order (year of start of study) for each province.

Location (Province)	Year of study	Study type	Study population	No. positive/no. examined (Prevalence)	Diagnostic Technique	Source/Sponsor
Kampong Cham Province	2000	Cross-sectional	School children (1 villages)	9/58 (15.5%)	FD	Sasakawa Foundation
	2012 (Dec)	Cross-sectional	Adults (5 villages)	157/658 (23.9%)	K-K, PCR Quasi-random	JSPS KAKENHI
	2013 (Sep)	Cross-sectional	Adults (8 villages)	224/533 (42.0%)	K-K, PCR	JSPS KAKENHI
Kampong Thom Province	2012 (Jun–Jul)	Cross-sectional	Adults (2 villages)	100/287 (34.8%)	K-K, PCR Quasi-random	JSPS KAKENHI
	2013 (Sep)	Cross-sectional	Adults (1 villages)	8/75 (10.7%)	K-K, PCR Quasi-random	JSPS KAKENHI
Kampot Province	2007	Cross-sectional	General Population (9 villages)	0/1,006 (0.0%)	K-K	ADB
Kandal Province	2012 (Dec)	Cross-sectional	Adults (9 villages)	22/518 (4.2%)	K-K Quasi-random	JSPS KAKENHI
Kratie Province	1996–97	Cross-sectional	General Population (19 villages)	450/2,500 (18.0%)	K-K	MSF
	1997–2010	Sentinel site surveillance	General Population (22 villages)	70/1,800 (3.9%)	K-K	WHO
	2000	Cross-sectional	School children (2 villages)	0/97 (0%)	FD	Sasakawa Foundation

(*Continued*)

Table 4 Identified Unpublished Primary Data on Human Opisthorchiasis/Small Fluke Infections in Cambodia. The studies are listed in chronological order (year of start of study) for each province.—cont'd

Location (Province)	Year of study	Study type	Study population	No. positive/no. examined (Prevalence)	Diagnostic Technique	Source/Sponsor
	2011	Cross-sectional	General Population (4 villages)	6/2,175 (0.3%)	K-K, FEC	IDRC
Modulkiri Province	2007	Cross-sectional	General Population (9 villages)	0/1,063 (0.0%)	K-K	ADB
Preah Vihear Province	2010	Cross-sectional	General Population (20 villages)	12/870 (1.4%)	FEC	UBSOF
	2010	Cross-sectional	General Population (1 villages)	32/70 (45.7%)	K-K	UBSOF
Prey Veng Province	2010	Cross-sectional	Adults (1 villages)	0/207 (0%)	K-K All household	Hiwa-Nakajima research Foundation
Ratanakiri Province	2007	Cross-sectional	General Population (9 villages)	7/1,049 (0.7%)	K-K	ADB
	2012 (Dec)	Cross-sectional	Adults (3 villages)	16/442 (3.6%)	K-K Quasi-random	JSPS KAKENHI

Stung Trang Province	2000	Cross-sectional	School children (2 villages)	0/94 (0%)	FD	Sasakawa Foundation
	2007	Cross-sectional	General Population (9 villages)	0/1,010 (0.0%)	K-K	ADB
Takeo Province	2005	Cross-sectional	Pregnant Women (9 communes)	6/190 (3.2%)	K-K	Toyota Foundation
	2005	Cross-sectional	School Children (8 schools)	0/437 (0%)	K-K	Toyota Foundation
	2005	Cross-sectional	School teachers (8 schools)	0/124 (0%)	K-K	Toyota Foundation
	2007	Cross-sectional	General Population (9 villages)	6/1,090 (0.6%)	K-K	ADB
Tbaung Kmom Province	2013 (Sep)	Cross-sectional	Adults (3 villages)	97/230 (33.0%)	K-K, PCR Quasi-random	JSPS KAKENHI
Countrywide	2013	Cross-sectional	School children	74/1,800 (4.1%)	K-K	USAID

–, no details provided; ADB, Asian Development Bank; CNM, National Centre for Parasitology, Entomology and Malaria Control, Phnom Penh; FD, Formalin-detergent sedimentation technique; FEC, Formalin-ether concentration technique; IDRC, International Development Research Centre, Canada; JSPS KAKENHI, Japanese Society for the Promotion of Science; K-K, Kato-Katz technique; MSFs, Médecins sans Frontières; PCR, Polymerase chain reaction; UBSOF, UBS Optimus Foundation; USAID, United States Agency for International Development; WHO, World Health Organization.
Information retrieved from CNM's Annual Reports 1997 to 2016.

During the following period of the *S. mekongi* sentinel site surveillance (1997–2010) in Kratie, the *O. viverrini* prevalence was merely 3.9% and an additional survey in 2011 revealed an even lower prevalence of 0.3%. All these surveys had rather high study samples of 1,800–2,500 individuals. Smaller surveys in 2000 in schoolchildren as part of the *S. mekongi* surveillance along the Mekong river provinces showed inconsistent *O. viverrini* prevalences of 0% (Kratie and Stung Treng provinces) and 15.5% (Kampong Cham province). Particularly the generally decreasing *O. viverrini* prevalence rates over time in Kratie province is also confirmed by published data (Sohn et al., 2012; Yong et al., 2014) and could reflect a positive side effect of the previously mentioned annual praziquantel mass-drug administration (MDA; see Section 2.2 above), which was implemented as control intervention alongside the *S. mekongi* sentinel site surveillance.

Unpublished prevalence data from Takeo province is uniformly low. Cross-sectional surveys in pregnant women, schoolchildren, and school teachers in 2005 resulted in prevalences between 0.0% and 3.2%. One community-wide study in 9 villages in 2007 revealed prevalences of 0.6%. These results are in line with some of the published data for 2006–07 (0.2% in adults) (Miyamoto et al., 2014) and 2011 (0.2% in the general population) (Khieu et al., 2014), but sharply contrast with others, which report prevalences ranging from 23.8% to 47.5% in the general population (Sohn et al., 2011b; Yong et al., 2012, 2014).

Unpublished and published data for Kampot province, Mondulkiri province, Ratanakiri province, Stung Treng province, and Prey Veng province consistently indicate low human *O. viverrini* prevalences. In contrast, the data for Kandal are more ambiguous with an unpublished report of 4.2% prevalence in adults in 2012 and published prevalences of 20.2% in adults in 2008 (Miyamoto et al., 2014), 7.9% in schoolchildren in 2009 (Khieu et al., 2013), and 5.7% also in schoolchildren in 2011 (Bless et al., 2015). Even more extreme, an unpublished study in the general population of 20 villages in Preah Vihear province in 2010, mainly aiming at protozoan infections and applying the formalin-ether concentration technique, diagnosed 1.4% of the individuals as *O. viverrini* positive. In contrast, a small exploratory survey in a suspected high-risk village (Kampong Samy village) – characterized by its location along the Mekong River and the widespread habit of eating undercooked/raw fish – in the same province (Preah Vihear) and year (2010) revealed a prevalence of 45.7% in the general population by using the Kato-Katz diagnostic technique. The only two identified published data points suggest prevalences of 2.7% (Yong et al., 2014) and

5.1% (Schär et al., 2014) in the years 2006–12 in the general population. For Kampong Cham – including the since 2014 separated province of Tbaung Kmom – 8 out of 9 unpublished and published data sources indicate high prevalences of 15.5%–58.4% and only one smaller survey by Lee and colleagues (Lee et al., 2002) in schoolchildren in 2002 using the formalin-ether concentration diagnostic technique found merely 4% positive for *O. viverrini* infections. Also Kampong Thom province seems to be highly endemic, with 3 out of 4 unpublished and published data sources indicating prevalences between 10.7% and 34.8% and only one published source (Yong et al., 2014) mentioning a low prevalence of 1.3% in the general population in 2006–11.

The only nation-wide survey – not in the general population but in school-children; however, resulted in an overall human opisthorchiasis prevalence rate of 4.1%.

4. DISCUSSION

To the best of our knowledge, this is the first time that information about *O. viverrini* infections in humans and animals from published and unpublished data are synthesised to examine this possible emerging public health problem in Cambodia.

First, it is important to note, that we did not find any report or publication on morbidity and/or mortality associated with *O. viverrini* infection in Cambodia. *O. viverrini* is a major risk factor for severe hepatobiliar morbidity, including cholangiocarcinoma, a fatal bile duct cancer (Edwards et al., 2018). It is of utmost importance to start investigation of hepatobiliar morbidity in *O. viverrini* endemic settings in Cambodia in order to assess the level morbidity burden linked with *O. viverrini* infection.

Even though only two major changes had to be made in the maps on human and animal opisthorchiasis (Figs. 2 and 3) due to the complementation of published with unpublished data (as explained in Section 3.2 above), the substantial value of checking locally available, thus far, unpublished sources is illustrated by the relative number of additionally identified data points (Table 4 in relation to Table 2).

In March 2006, a high infection prevalence of 26.9% *O. viverrini* infection was detected in Angsvay Cheak village, Takeo province (Miyamoto et al., 2014). This observation was turning point after which specific *O. viverrini* studies were launched by CNM in collaboration with Japanese and Korean researcher teams with the objective to document liver fluke and other intestinal trematode transmission in Cambodia. Information on *O. viverrini* before that date was rather incidental findings of soil-transmitted helminth or schistosomiasis surveillance surveys.

Overall, despite all efforts to identify as many data sources as possible, a rather uncomplete and very spotty pattern emerges from the data. On the one hand, this is probably a reflection of the diversity of the various settings in terms of ecology, infrastructure, socio-economic wealth, eating habits, etc., and the known fact that these factors usually lead to a highly focal occurrence of opisthorchiasis. The revealed pattern that some of the provinces along the big Mekong and Tonle Sap Rivers have lower rates than other non-bordering provinces may also indicate that the transmission is particularly frequent in the vicinity of smaller rivers, ponds and rice fields of the most rural areas of Cambodia, where the habit of raw or insufficiently cooked fish consumption is very common. Indeed, in the fast flowing and water-rich Mekong and Tonle Sap Rivers the transmission of the parasite from human to molluscs and further to fish is hampered by the water velocity and quantity. On the other hand, the mapped pattern is almost certainly also an artefact produced by the lack of data. In many provinces in western Cambodia only zero or one data points was found (see Tables 2–4, Figs. 2–3). Furthermore, due to repeatedly missing meta-data on exact locations, the here provided maps could only be drawn at provincial level (Figs. 2 and 3). Having also the previously mentioned focal occurrence of opisthorchiasis in mind, maps at higher geographical resolution are an absolute necessity for better understanding the *O. viverrini* epidemiology and endemicity in Cambodia and guiding national control efforts.

Most of the identified studies utilised the K-K technique, which does not allow for the differentiation between the eggs of *O. viverrini, C. sinensis*, and other small fluke eggs (e.g. species of the families *Gymnophallidae, Heterophyidae, Lecithodendriidae*, or *Plagiorchiidae*; see also Appendix 1, Table A1, for an overview over the taxonomy of food-borne trematodes) (Johansen et al., 2010; Fürst et al., 2012c; Wongsawad and Wongsawad, 2012; Clausen and Murrell, 2014; Maguire, 2015). However, two references report on the results of studies using PCR on stool samples and purging of patients to retrieve also adult trematode worms and both definitely confirm the presence of *O. viverrini* in Cambodia (Sohn et al., 2012; Miyamoto et al., 2014). It is important to note, that both studies confirmed the presence of

O. viverrini and both did not report *C. sinensis*. This is consistent with other studies, which indicate a clear separation between the endemic areas of *O. viverrini* (Thailand, Cambodia, Laos, southern Vietnam) and *C. sinensis* (China, Korea, northern Vietnam) (Rim, 2005; International Agency for Research on Cancer, 2012; Sithithaworn et al., 2012a). Nevertheless, a publication from 2009 speculates about *C. sinensis* potentially extending its range also to central Thailand (Traub et al., 2009; Sripa et al., 2010). Furthermore, *Opisthorchis lobatus* has been recently discovered as a new small fluke infection in fish in the Greater Mekong sub-region and it has been emphasised that "questions regarding human infection and the extent of the geographic distribution of these species should be investigated" (Thaenkham et al., 2011). Hence, the distribution of *Opisthorchis* spp., *C. sinensis* and other small flukes in the Greater Mekong sub-region seems to be more heterogeneous than previously anticipated and the respective levels of endemicities need further investigation with highly sensitive diagnostic techniques that allow for the clear differentiation of the parasite species in humans and animals.

With regard to the animal intermediate hosts, it is well-known that *Bithynia* snails are present in many areas of the Lower Mekong region, including Cambodia, which is also well-recognised as a biodiversity hotspot for gastropod molluscs (Petney et al., 2012, 2013). Even though the data on *O. viverrini* infections in snails is very limited, there is at least one reference reporting on findings from three Cambodian provinces (Fig. 3, Table 3) (Miyamoto et al., 2014). Besides identified data on natural *O. viverrini* infections in various secondary intermediate host fish species (Table 3), also experimental infections in widely reared and consumed fish species demonstrate the many opportunities for the parasite to complete its life cycle in the region and to ultimately infect humans (Donthaisong et al., 2014). Of note, one of the seven provinces, where data on fish infections were reported, namely Prey Veng, belongs to the only three provinces, where among examined people no infection in humans (Figs. 2 and 3). However, it must be said that humans were examined only in two villages. As the parasite is present in secondary intermediate host fish in Prey Veng province, it is also very likely that human infections occur there and that the search for human infection was missing the affected village/population group. The only two provinces that remain without reports on either human or animal infections are Kampong Speu and Kep in the very South.

Also remarkable with regard to animal infections is the fact that the identified values for prevalence and number of cercariae in snail infections are considerably lower than the same values in fish (Table 3). This finding reflects the common feature of the trematode biology that only very few infected first intermediate hosts are sufficient to infect a much bigger number of secondary intermediate hosts and to sustain the parasites' life cycle (Vonghachack et al., 2017). Furthermore, it is noteworthy that cats, dogs, and more rarely also other fish eating mammals are known as *O. viverrini* reservoir hosts, but no data could be found on this aspect in Cambodia. Even though it currently seems that humans may play the major role in disease transmission as suggested also by a recent mathematical modelling study based on data from Laos (Bürli et al., 2018), reservoir hosts may still influence the disease dynamics and should therefore receive further attention (World Health Organization, 1995; Petney et al., 2013).

Indications for the *O. viverrini* endemicity of Cambodia also come from the many case reports of infections in Cambodian refugees and immigrants, which date back to the early 1980s (Ambroise-Thomas et al., 1980, 1984, 1985; Löscher et al., 1981; Tateno et al., 1981; Bouree et al., 1982; Keittivuti et al., 1982, 1983; Vernes et al., 1982; Luong Dinh Giap et al., 1983; Parish, 1985; Yangco et al., 1987; Gyorkos et al., 1992; Stauffer et al., 2003, 2004). Interestingly, two studies from North America noted that the opisthorchiasis prevalences in Cambodian refugees remained virtually unchanged at ca. 4—5% even six years after resettlement (Gyorkos et al., 1992) and that the refugees maintained strong ties to their traditional diets and native foods for years (Story and Harris, 1989), which highlights the importance and inertia of deeply rooted behavioural patterns. Furthermore, evidence from additional studies on other human food-borne trematode infections caused by the ingestion of raw or undercooked food of aquatic origin (Im et al., 1997; Park et al., 2004; Sohn et al., 2011c; Chai, 2013) highlight the existence of the risky behaviour of consuming such food among Cambodians. While it has been hypothesised in the past that the consumption of raw fish is not a habit in Khmer communities and that opisthorchiasis transmission via this route may therefore be unlikely (Stich et al., 1999; Grundy-Warr et al., 2012), recent studies mention some frequently used Cambodian food ingredients (i.e. *prahoc* — Khmer fermented fish paste), which are made of raw fish, may be linked to the transmission of opisthorchiasis and other small fluke infections (Grundy-Warr et al., 2012;

Clausen and Murrell, 2014; Sithithaworn et al., 2014) In addition, it has been noted that potential transmission pathways could arise from cross-contamination of kitchen utensils (e.g. knifes, chopping boards) and their repeated use without proper cleaning during unsafe food preparation (Grundy-Warr et al., 2012; Clausen and Murrell, 2014).

There are many additional factors, which need to be considered, in the assessment of the level of endemicity and whether *O. viverrini* is emerging in Cambodia or not. One important factor is that "average fish-dependence of Cambodians is the highest of all countries in the Mekong Basin, and [that] the Mekong River and Tonle Sap Lake contain many *Cyprinid* fish species that are also found in opisthorchiasis-endemic Thailand and southern Laos" (Grundy-Warr et al., 2012). Additionally, aquacultural food production is steadily increasing in Cambodia and the Southeast Asian region, and its importance as a supply for affordable animal protein for large parts of the population as well as a source of income is unquestioned (World Health Organization, 1995; World Health Organization and Food and Agriculture Organization, 2004; Keiser and Utzinger, 2005; Hortle, 2008; Keiser and Utzinger, 2009; Ziegler et al., 2013; Food and Agriculture Organization, 2015). Concerns have been expressed that the planned construction of a large number of hydropower dams along the Mekong and its tributaries may reduce the naturally occurring fish diversity and abundance, leading to a substitution by fish production in water reservoirs behind the dams and in small-scale ponds (Ziegler et al., 2013). Unfortunately, some examples indicate that these production methods may facilitate the occurrence of the full life cycle of *O. viverrini* and other small flukes and this threat may be escalated in particular by an often unsafe sanitary situation, the use of human excreta as fertilizer in the agriculture, the attempt of nutrient recycling in waste-fed aquaculture, the general lack of knowledge and food inspections, the national and international trade among fish farms, the general livestock and food trade, and human population growth and migration (Petney, 2001; World Health Organization and Food and Agriculture Organization, 2004; Keiser and Utzinger, 2005; Hortle, 2008; Keiser and Utzinger, 2009; Coker et al., 2011; Conlan et al., 2011; Lima dos Santos and Howgate, 2011; Sithithaworn et al., 2012b; Petney et al., 2013; Piraksakulrat et al., 2013; Ziegler et al., 2013; Sithithaworn et al., 2014).

Climate change is often said to cause more flooding in Southeast Asia, which may not only extend the habitats for the parasites' intermediate hosts, but also increase the environmental contamination with human and animal faeces containing the parasites' eggs. Furthermore, predicted warmer surface air temperatures may increase cercarial production rates in the intermediate host snails. All these effects would enhance *O. viverrini* and other small fluke abundance and transmission (Yang et al., 2010; Conlan et al., 2011). However, there are also some climate change models, which predict decreased precipitation and consequently longer periods of droughts in the Lower Mekong area leading to dramatic intermediate host population crashes and ultimately a reduction in *O. viverrini* and other small fluke abundance and transmission (Selstad Utaaker and Robertson, 2015). Overall, the situation with regard to climate change is somewhat inconclusive and speculative (Petney et al., 2013) and it is also worth mentioning that in some areas, where *O. viverrini* infections have diminished, other small fluke infections seem to be now prevalent (Lima dos Santos and Howgate, 2011).

Despite the compelling evidence of Cambodia as an endemic country for opisthorchiasis and other small fluke infections, there were hardly any reliable data on these infections at a national level until very recently. An early review published in 1968 reported "*Opisthorchis sinensis*" as the only trematode in humans in Cambodia and otherwise only large trematode infections in animals (Segal et al., 1968). Later WHO reports were unable to provide more detailed information (World Health Organization, 1995; World Health Organization and Food and Agriculture Organization, 2004; World Health Organization, 2008) and current burden of disease estimation projects basically predict no cases of opisthorchiasis in Cambodia because of the unavailability of suitable data (Fürst et al., 2012b; Torgerson et al., 2015; GBD DALYs and HALE Collaborators, 2017). The level of uncertainty is also reflected by anecdotal evidence of human opisthorchiasis prevalence of up to 40% in Kratie province (Conlan et al., 2011), which contrasts with most identified primary data for this province of only 3–5% (Tables 2 and 4). However, two WHO reports from 2009 to 2011 clearly mention that *O. viverrini* has to be considered endemic in Cambodia and that mapping has been initiated with reported prevalences ranging between 1% and 25%, ongoing transmission in seven provinces, and an estimated population at risk of 560,000 (World Health Organization, 2009, 2011). Unfortunately, these reports also state that the mapping efforts had to be stopped due to funding constraints and emphasise at the same time the need for additional surveys (World Health Organization, 2009; World Health Organization, 2011). The national plan for the control of neglected tropical diseases in Cambodia reported that prevalences range between 0.1% and 54% in 14 provinces (National Centre for Parasitology Entomology and Malaria Control, 2011) and all 14 provinces are also listed in Tables 2 and/or 4. In 2012, the International Agency for Research on Cancer estimated the national prevalence of opisthorchiasis in humans at 2–5% and reported that 600,000 people are infected with *O. viverrini* in Cambodia (International Agency for Research on Cancer, 2012). However, some experts consider even this figures as a marked underestimate and stress the need for further studies (Sithithaworn et al., 2012a).

Having all these data in mind, the human opisthorchiasis prevalence of 4.1% from the only identified unpublished nation-wide survey in school-children conducted in 2013 (Table 4) may currently be a good starting point to approximate the opisthorchiasis prevalence at the national level. It has been noted that, even though the total *O. viverrini* prevalences in humans may widely vary from location to location, the age-specific prevalence patterns in a population are usually similar. The human opisthorchiasis prevalence rates increase with increasing age in children and adolescents and reach a plateau in adults (Sithithaworn et al., 2007). Applying previously published generalized age-prevalence patterns (Fürst et al., 2012b) on the 4.1% prevalence in school-children, assuming that these school-children are between 5 and 14 years old, and using United Nations' age- and sex-specific population estimates for 2015 for Cambodia (United Nations Population Division, 2017), the corrected prevalence in the total population would equal 4.9%. This prevalence would correspond to about 753,000 people infected in Cambodia in total.

Helminth control in Cambodia is overseen by the National Center for Parasitology, Entomology and Malaria Control and control activities are pursued in collaboration with several other actors (World Health Organization, 2009; National Centre for Parasitology Entomology and Malaria Control, 2011). In the past, routine control activities for opisthorchiasis was merely comprised of accidental treatment with praziquantel in the frame of schistosomiasis control (Stich et al., 1999; Chigusa et al., 2006; Muth, 2006; Muth et al., 2007, 2010; World Health Organization, 2015) and it was estimated that only a very small proportion of opisthorchiasis patients received praziquantel this way (Montresor et al., 2008; National

Centre for Parasitology Entomology and Malaria Control, 2011). More recently, an opisthorchiasis control strategy was developed, which plans annual mass drug administration (MDA) with praziquantel to the entire population in districts with opisthorchiasis prevalences of >20%, MDA every two years in districts with prevalences of 5–20%, and individual case management of suspected opisthorchiasis patients reporting to health facilities (National Centre for Parasitology Entomology and Malaria Control, 2011; World Health Organization, 2011). Also more integrated activities are envisaged such as health education on hygienic eating practices and agricultural production processes, collaborations with the animal health and food safety sectors, and monitoring, evaluation and operational research activities (National Centre for Parasitology Entomology and Malaria Control, 2011). However, the national plan for the control of neglected tropical diseases 2011–15 also mentions that "passive case detection and treatment has been in place in a few places, but mass treatment strategies have not been implemented due to many reasons" (National Centre for Parasitology Entomology and Malaria Control, 2011). Furthermore, and due to the aforementioned funding gaps and partly still missing information on the extent of the problem, the planned opisthorchiasis control interventions are likely to be insufficient. Further useful stimuli could certainly come from the Regional Network on Asian Schistosomiasis and Other Helminth Zoonoses (RNAS+), which aims at strengthening research capabilities and also closing the gap between research and disease control in truly integrated approaches (Olveda et al., 2010; Yang et al., 2010). Cambodia is a member country of the RNAS+ and food-borne trematodiasis is one the RNAS+ -work nodules (Olveda et al., 2010; Yang et al., 2010).

We have performed a thorough literature review on existing data on *O. viverrini* infection. For that purpose, we have searched five independent international databases and reviewed reference lists of key publications. Furthermore, we have searched unpublished reports from various sources. Despite these extensive efforts it might be possible that not all information was found particularly those published in Chinese, Russian or Thai languages. Furthermore, we did not undertake a quality check of the identified data source which might have resulted in an inappropriate combination of studies and reports. Also, we did not correct the observed prevalence rates with the sensitivity and specificity of the diagnostic techniques and we did not adjust for age and gender of the examined population. These short comings might have lead in an underestimation of the true infection rates and distribution.

Nevertheless our results show that (i) a country-wide assessment of the importance of *O. viverrini* infection; (ii) our collected information indicate a North-South belt of central provinces with higher average reported prevalence rates but in general draw a patchy distribution of infection across Cambodia where it is not clear in how far the missing data in most locations is responsible for this distribution or the highly focal transmission of *O. viverrini*; (iii) in some examined areas the infection rates exceed 20% of the general population which warrants a public health intervention such as MDAs and IEC campaigns; (iv) infection in fish and molluscs populations could be documented which indicate on-going transmission; (v) among studies where liver fluke species was identified, only *O. viverrini* was found (though, it cannot be excluded that other species are also endemic); and (vi) no information on morbidity and mortality linked with liver fluke infection could be found.

Considering all aforementioned evidence, it seems plausible that *O. viverrini* and other small fluke infections were — and still are — underreported in Cambodia. Furthermore, it is likely that the real abundance of the food-borne trematodiasis and their transmission will further increase in the future with serious consequences for human health — including an increase in highly fatal liver fluke associated bile duct cancers (Sripa et al., 2012) — and with an unclear impact on the country's socioeconomic development. The National Helminth Control Program has the policy ready to implement a thorough liver fluke control program. However, coherent data on infection and resulting disease across Cambodia is required to guide these control efforts. A national survey, concerted specific surveys, and continued surveillance can be immediate measures to facilitate the data collection required for the planning of adequate interventions.

ACKNOWLEDGEMENTS

We would like to thank the Swiss Tropical and Public Health Institute's library team (Rebekka Hirsbrunner, Hanna Walter, Tenzin Shagatsang, Giovanni Casagrande) for their fantastic support and the organization even of almost untraceable manuscripts. TF is grateful to the Swiss National Science Foundation (SNSF) for an Advanced Postdoc Mobility fellowship and a Return Grant (project nos. P300P3-154634/1 and P300P3-154634/2). The funders had no role in the conceptualisation, analysis and publication of this article.

APPENDIX 1

Table A1 Food-borne Trematodes' Taxonomy, Human Diseases Caused, and the Respective Source of Transmission.

Order	Family	Species	Fluke type	Disease	Transmission source
(Digenea)	Neodiplostomidae	*Fibricola cratera*	Intestinal fluke	Neodiplostomiasis	Food-borne: frog, snake
		Neodiplostomum seoulense	Intestinal fluke	Neodiplostomiasis	Food-borne: frog, snake, tadpole
Echinostomida	Cathaemaciidae	*Cathaemacia cabrerai*	Intestinal fluke	Echinostomiasis	Food-borne: unknown
	Echinostomatidae	*Acanthoparyphium tyosenense*	Intestinal fluke	Echinostomiasis	Food-borne: mussel, snail
		Artyfechinostomum malayanum	Intestinal fluke	Echinostomiasis	Food-borne: snail
		Artyfechinostomum oraoni	Intestinal fluke	Echinostomiasis	Food-borne: unknown
		Echinochasmus fujianensis	Intestinal fluke	Echinostomiasis	Food-borne: fish
		Echinochasmus japonicus	Intestinal fluke	Echinostomiasis	Food-borne: fish
		Echinochasmus jiufoensis	Intestinal fluke	Echinostomiasis	Food-borne: unknown
		Echinochasmus liliputanus	Intestinal fluke	Echinostomiasis	Food-borne: fish
		Echinochasmus perfoliatus	Intestinal fluke	Echinostomiasis	Food-borne: fish
		Echinoparyphium recurvatum	Intestinal fluke	Echinostomiasis	Food-borne: snail
		Echinostoma angustitestis	Intestinal fluke	Echinostomiasis	Food-borne: fish
		Echinostoma cinetorchis	Intestinal fluke	Echinostomiasis	Food-borne: fish, snail
		Echinostoma echinatum	Intestinal fluke	Echinostomiasis	Food-borne: mussel, snail
		Echinostoma hortense	Intestinal fluke	Echinostomiasis	Food-borne: fish
		Echinostoma ilocanum	Intestinal fluke	Echinostomiasis	Food-borne: snail
		Echinostoma macrorchis	Intestinal fluke	Echinostomiasis	Food-borne: snail
		Echinostoma malayanum	Intestinal fluke	Echinostomiasis	Food-borne: snail
		Echinostoma revolutum	Intestinal fluke	Echinostomiasis	Food-borne: mussel, snail
		Episthmium caninum	Intestinal fluke	Echinostomiasis	Food-borne: fish
		Himasthla muehlensi	Intestinal fluke	Echinostomiasis	Food-borne: mussel
		Hypoderaeum conoideum	Intestinal fluke	Echinostomiasis	Food-borne: snail
		Isthmiophora melis	Intestinal fluke	Echinostomiasis	Food-borne: tadpole
		Psilorchis hominis	Intestinal fluke	Echinostomiasis	Food-borne: unknown

	Fasciolidae	*Fasciola gigantica*	Liver fluke	Fascioliasis	Food-borne: plant, maybe water
		Fasciola hepatica	Liver fluke	Fascioliasis	Food-borne: plant, maybe water
		Fasciolopsis buski	Intestinal fluke	Fasciolopsiasis	Food-borne: plant
	Gastrodiscidae	*Gastrodiscoides hominis*	Intestinal fluke	Echinostomiasis	Food-borne: crab, frog, plant, tadpole
	Paramphistomidae	*Fischoederius elongatus*	Intestinal fluke	Echinostomiasis	Food-borne: plant
		Watsonius watsoni	Intestinal fluke	Echinostomiasis	Food-borne: plant
Opisthorchiida	Heterophyidae	*Apophallus donicus*	Intestinal fluke	Heterophyiasis	Food-borne: fish
		Ascocotyle longa	Intestinal fluke	Heterophyiasis	Food-borne: fish
		Centrocestus armatus	Intestinal fluke	Heterophyiasis	Food-borne: fish
		Centrocestus caninus	Intestinal fluke	Heterophyiasis	Food-borne: fish
		Centrocestus cuspidatus	Intestinal fluke	Heterophyiasis	Food-borne: fish
		Centrocestus formosanus	Intestinal fluke	Heterophyiasis	Food-borne: fish
		Centrocestus kurokawai	Intestinal fluke	Heterophyiasis	Food-borne: unknown
		Cryptocotyle lingue	Intestinal fluke	Heterophyiasis	Food-borne: fish
		Haplorchis pleurolophocerca	Intestinal fluke	Heterophyiasis	Food-borne: fish
		Haplorchis pumilio	Intestinal fluke	Heterophyiasis	Food-borne: fish
		Haplorchis taichui	Intestinal fluke	Heterophyiasis	Food-borne: fish
		Haplorchis vanissimus	Intestinal fluke	Heterophyiasis	Food-borne: fish
		Haplorchis yokogawai	Intestinal fluke	Heterophyiasis	Food-borne: fish
		Heterophyes dispar	Intestinal fluke	Heterophyiasis	Food-borne: fish
		Heterophyes heterophyes	Intestinal fluke	Heterophyiasis	Food-borne: fish
		Heterophyes nocens	Intestinal fluke	Heterophyiasis	Food-borne: fish
		Heterophyopsis continua	Intestinal fluke	Heterophyiasis	Food-borne: fish
		Metagonimus minutus	Intestinal fluke	Heterophyiasis	Food-borne: fish

(*Continued*)

Table A1 Food-borne Trematodes' Taxonomy, Human Diseases Caused, and the Respective Source of Transmission.—cont'd

		Metagonimus miyatai	Intestinal fluke	Heterophyiasis	Food-borne: fish
		Metagonimus takahashii	Intestinal fluke	Heterophyiasis	Food-borne: fish
		Metagonimus yokogawai	Intestinal fluke	Heterophyiasis	Food-borne: fish
		Procerovum calderoni	Intestinal fluke	Heterophyiasis	Food-borne: fish
		Procerovum varium	Intestinal fluke	Heterophyiasis	Food-borne: unknown
		Pygidiopsis summa	Intestinal fluke	Heterophyiasis	Food-borne: fish
		Stellantchasmus falcatus	Intestinal fluke	Heterophyiasis	Food-borne: fish
		Stellantchasmus formosanus	Intestinal fluke	Heterophyiasis	Food-borne: fish
		Stellantchasmus pseudocirratus	Intestinal fluke	Heterophyiasis	Food-borne: fish
		Stictodora fuscata	Intestinal fluke	Heterophyiasis	Food-borne: fish
		Stictodora lari	Intestinal fluke	Heterophyiasis	Food-borne: fish
	Opisthorchiidae	*Clonorchis sinensis*	Liver fluke	Clonorchiasis	Food-borne: fish
		Opisthorchis felineus	Liver fluke	Opisthorchiasis	Food-borne: fish
		Opisthorchis viverrini	Liver fluke	Opisthorchiasis	Food-borne: fish
Plagiorchiida	Dicrocoeliidae	*Dicrocoelium dendriticum*	Liver fluke	Dicrocoeliasis	Food-borne: ant
	Lecithodendriidae	*Phaneropsolus bonnei*	Intestinal fluke	Lecithodendriidiasis	Food-borne: fly, naiad of fly
		Phaneropsolus spinicirrus	Intestinal fluke	Lecithodendriidiasis	Food-borne: unknown
		Prosthodendrium molenkampi	Intestinal fluke	Lecithodendriidiasis	Food-borne: fly, naiad of fly
	Microphallidae	*Spelotrema brevicaeca*	Intestinal fluke	Microphallidiasis	Food-borne: crab, shrimp
	Nanophyetidae	*Nanophyetus salmincola*	Intestinal fluke	Nanophyetiasis	Food-borne: fish
	Paragonimidae	*Paragonimus africanus*	Lung fluke	Paragonimiasis	Food-borne: crab, shrimp, mammal meat
		Paragonimus heterotremus	Lung fluke	Paragonimiasis	Food-borne: crab, shrimp, mammal meat

		Paragonimus kellicotti	Lung fluke	Paragonimiasis	Food-borne: crab, shrimp, mammal meat
		Paragonimus mexicanus	Lung fluke	Paragonimiasis	Food-borne: crab, shrimp, mammal meat
		Paragonimus skrjabini (including subspecies)	Lung fluke	Paragonimiasis	Food-borne: crab, shrimp, mammal meat
		Paragonimus uterobilateralis	Lung fluke	Paragonimiasis	Food-borne: crab, shrimp, mammal meat
		Paragonimus westermani	Lung fluke	Paragonimiasis	Food-borne: crab, shrimp, mammal meat
	Plagiorchiidae/	*Plagiorchis harinasutai*	Intestinal fluke	Plagiochiasis	Food-borne: unknown
		Plagiorchis javensis	Intestinal fluke	Plagiochiasis	Food-borne: insect larva
		Plagiorchis muris	Intestinal fluke	Plagiochiasis	Food-borne: fish, insect, snail
		Plagiorchis philippinensis	Intestinal fluke	Plagiochiasis	Food-borne: insect larva
Strigeata	Brachylaimidae	*Brachylaima cribbi*	Intestinal fluke	Brachylaimiasis	Food-borne: snail
	Gymnophallidae	*Gymnophalloides seoi*	Intestinal fluke	Gymnophalloidiasis	Food-borne: mussel
Strigeatida	Strigeidae	*Cotylurus japonicus*	Intestinal fluke	Strigeidiasis	Food-borne: snail

The table is based on references Blair et al., 2007; Chai, 2007; Mas-Coma et al., 2007; Sithithaworn et al., 2007 and has been previously published as a webappendix in Fürst et al., 2012b.

REFERENCES

Ambroise-Thomas, P., Goullier, A., Peyron, F., 1984. Therapeutic results in opisthorchiasis with praziquantel in a reinfection-free environment in France. Arzneimittelforschung 34, 1177–1179.

Ambroise-Thomas, P., Peyron, F., Goullier, A., Wegner, D.H., 1985. Praziquantel in the treatment of opisthorchiasis in Southeast Asian refugees. Evaluation of 153 cases. Bull. Soc. Pathol. Exot. Filiales 78, 492–499.

Ambroise-Thomas, P., Wegner, D.H., Goullier, A., Desgeorges, P.T., Billiault, X., 1980. Clinical trial of niclofolan (Bay 9015) in the treatment of human opisthorchiasis. Apropos of 51 cases. Bull. Soc. Pathol. Exot. Filiales 73, 293–301.

Blair, D., Agatsuma, T., Wang, W., 2007. Paragonimiasis. In: Murrell, K.D., Fried, B. (Eds.), World Class Parasites, Food-borne Parasitic Zoonoses - Fish and Plant-Borne Parasites, vol. 11. Springer, New York (USA), pp. 117–150.

Bless, P.J., Schär, F., Khieu, V., Kramme, S., Muth, S., Marti, H., et al., 2015. High prevalence of large trematode eggs in schoolchildren in Cambodia. Acta Trop. 141, 295–302.

Bouree, P., Thulliez, P., Taugourdeau, P., 1982. Southeast Asian refugees parasitic infections. Mol. Biochem. Parasitol. (Suppl. 1), 289–290.

Bürli, C., Harbrecht, H., Odermatt, P., Sayasone, S., Chitnis, N., 2018. Mathematical analysis of the transmission dynamics of the liver fluke, *Opisthorchis viverrini*. J. Theor. Biol. 439, 181–194.

Chai, J.-Y., 2007. Intestinal flukes. In: Murrell, K.D., Fried, B. (Eds.), World Class Parasites, Food-borne Parasitic Zoonoses - Fish and Plant-Borne Parasites, vol. 11. Springer, New York (USA), pp. 53–115.

Chai, J.-Y., 2013. Paragonimiasis. In: Garcia, H.H., Tanowitz, H.B., Del Brutto, O.H. (Eds.), Handbook of Clinical Neurology, vol. 114. Elsevier, Amsterdam (Netherlands), pp. 283–296 (Chapter 23).

Chai, J.-Y., Sohn, W.-M., Na, B.-K., Yong, T.-S., Eom, K.S., Yoon, C.-H., et al., 2014. Zoonotic trematode metacercariae in fish from Phnom Penh and Pursat, Cambodia. Kor. J. Parasitol. 52, 35–40.

Chigusa, Y., Ohmae, H., Otake, H., Keang, H., Sinuon, M., Saem, C., et al., 2006. Effects of repeated praziquantel treatment on schistosomiasis mekongi morbidity as detected by ultrasonography. Parasitol. Int. 55, 261–265.

Clausen, J.H., Murrell, D., 2014. Heminth - nematode: *Haplorchis*. In: Motarjemi, Y. (Ed.), Encyclopedia of Food Safety. Academic Press, Waltham (USA), pp. 99–103.

Coker, R.J., Hunter, B.M., Rudge, J.W., Liverani, M., Hanvoravongchai, P., 2011. Emerging infectious diseases in southeast Asia: regional challenges to control. Lancet 377, 599–609.

Conlan, J.V., Sripa, B., Attwood, S., Newton, P.N., 2011. A review of parasitic zoonoses in a changing Southeast Asia. Vet. Parasitol. 182, 22–40.

Donthaisong, C., Arunsan, P., Suwannatrai, K., Prasopdee, S., Kulsantiwong, J., Wongmaneeprateep, S., et al., 2014. Experimental infection of *Opisthorchis viverrini* cercariae to the cyprinid fish, *Barbonymus gonionotus*. Acta Trop. 136, 118–122.

Edwards, S.W., Spofford, E.M., Price, C., Wright, H.L., Salao, K., Suttiprapa, S., et al., 2018. Opisthorchiasis-induced cholangiocarcinoma: how innate immunity may cause cancer. Adv. Parasitol. 101, 149–176.

Food and Agriculture Organization, 2015. Fishery Statistical Collections - Global Aquaculture Production. Retrieved from: http://www.fao.org/fishery/statistics/global-aquaculture-production/en.

Fürst, T., Duthaler, U., Sripa, B., Utzinger, J., Keiser, J., 2012a. Trematode infections: liver and lung flukes. Infect. Dis. Clin. N. Am. 26, 399–419.

Fürst, T., Keiser, J., Utzinger, J., 2012b. Global burden of human food-borne trematodiasis: a systematic review and meta-analysis. Lancet Infect. Dis. 12, 210–221.
Fürst, T., Sayasone, S., Odermatt, P., Keiser, J., Utzinger, J., 2012c. Manifestation, diagnosis, and management of foodborne trematodiasis. BMJ 344, e4093.
GBD DALYs, HALE Collaborators, 2017. Global, regional, and national disability-adjusted life-years (DALYs) for 333 diseases and injuries and healthy life expectancy (HALE) for 195 countries and territories, 1990-2016: a systematic analysis for the Global Burden of Disease Study 2016. Lancet 390, 1260–1344.
Grundy-Warr, C., Andrews, R.H., Sithithaworn, P., Petney, T.N., Sripa, B., Laithavewat, L., et al., 2012. Raw attitudes, wetland cultures, life-cycles: socio-cultural dynamics relating to *Opisthorchis viverrini* in the Mekong Basin. Parasitol. Int. 61, 65–70.
Gyorkos, T.W., MacLean, J.D., Viens, P., Chheang, C., Kokoskin-Nelson, E., 1992. Intestinal parasite infection in the Kampuchean refugee population 6 years after resettlement in Canada. J. Infect. Dis. 166, 413–417.
Hortle, K.G., 2008. Liver and intestinal flukes: an underrated health risk in the Mekong Basin. Catch Culture 14, 14–18.
Im, J.-G., Chang, K.-H., Reeder, M.M., 1997. Current diagnostic imaging of pulmonary and cerebral paragonimiasis, with pathological correlation. Semin. Roentgenol. 32, 301–324.
International Agency for Research on Cancer, 2012. IARC Monographs on the Evaluation of Carcinogenic Risks to Humans. A Review of Carcinogen—Part B: Biological Agents. IARC, Lyon (France).
Johansen, M.V., Sithithaworn, P., Bergquist, R., Utzinger, J., 2010. Towards improved diagnosis of zoonotic trematode infections in Southeast Asia. Adv. Parasitol. 73, 171–195.
Kaewkes, S., 2003. Taxonomy and biology of liver flukes. Acta Trop. 88, 177–186.
Katz, N., Chaves, A., Pellegrino, J., 1972. A simple device for quantitative stool thick-smear technique in *Schistosomiasis mansoni*. Rev. Inst. Med. Trop. Sao Paulo 14, 397–400.
Keiser, J., Utzinger, J., 2005. Emerging foodborne trematodiasis. Emerg. Infect. Dis. 11, 1507–1514.
Keiser, J., Utzinger, J., 2009. Food-borne trematodiases. Clin. Microbiol. Rev. 22, 466–483.
Keittivuti, B., D'Agnes, T., Keittivuti, A., Viravaidya, M., 1982. Prevalence of schistosomiasis and other parasitic diseases among Cambodian refugees residing in Bang-Kaeng holding center, Prachinburi Province, Thailand. Am. J. Trop. Med. Hyg. 31, 988–990.
Keittivuti, B., Keittivuti, A., O'Rourke, T.F., 1983. Parasitic diseases with emphasis on schistosomiasis in Cambodian refugees in Prachinburi province, Thailand. Southeast Asian J. Trop. Med. Publ. Health 14, 491–494.
Khieu, V., Schär, F., Marti, H., Bless, P.J., Char, M.C., Muth, S., et al., 2014. Prevalence and risk factors of *Strongyloides stercoralis* in Takeo province, Cambodia. Parasites Vectors 7, 221.
Khieu, V., Schär, F., Marti, H., Sayasone, S., Duong, S., Muth, S., et al., 2013. Diagnosis, treatment and risk factors of *Strongyloides stercoralis* in schoolchildren in Cambodia. PLoS Neglect. Trop. Dis. 7, e2035.
Kobayashi, J., Chinen, K., Samidt, S., Higa, K., Chinen, M., Sato, Y., et al., 1999. Prevalence of intestinal parasite infection in Toul Roka village, Phnom Penh municipality, Cambodia. Jpn. J. Trop. Med. Hyg. 27, 517–519.
Lee, K.-J., Bae, Y.-T., Kim, D.-H., Deung, Y.-K., Ryang, Y.-S., Kim, H.-J., et al., 2002. Status of intestinal parasites infection among primary school children in Kampongcham, Cambodia. Kor. J. Parasitol. 40, 153–155.
Lima dos Santos, C.A.M., Howgate, P., 2011. Fishborne zoonotic parasites and aquaculture: a review. Aquaculture 318, 253–261.

Löscher, T., Nothdurft, H.D., Prüfer, L., von Sonnenburg, F., Lang, W., 1981. Praziquantel in clonorchiasis and opisthorchiasis. Tropenmed. Parasitol. 32, 234–236.
Luong Dinh Giap, G., Lam Tan, B., Faucher, P., Roche, M.C., Ripert, C., 1983. Hepatic distomatosis caused by *Clonorchis/Opisthorchis* spp. in refugees from South-East Asia. Effects of treatment with praziquantel. Med. Trop. 43, 325–330.
Maguire, J.H., 2015. Trematodes (schistosomes and liver, intestinal, and lung flukes). In: Bennett, J.E., Dolin, R., Blaser, M.J.M. (Eds.), Douglas, and Bennett's Principles and Practice of Infectious Diseases, eighth ed. Elsevier Saunders, Philadelphia (USA), pp. 3216–3226.
Mairiang, E., Mairiang, P., 2003. Clinical manifestation of opisthorchiasis and treatment. Acta Trop. 88, 221–227.
Maleewong, W., Intapan, P., Wongwajana, S., Sitthithaworn, P., Pipitgool, V., Wongkham, C., et al., 1992. Prevalence and intensity of *Opisthorchis viverrini* in rural community near the Mekong River on the Thai-Laos border in northeast Thailand. J. Med. Assoc. Thai. 75, 231–235.
Mas-Coma, S., Bargues, M.D., Valero, M.A., 2007. Plant-borne trematode zoonoses: fascioliasis and fasciolopsiasis. In: Murrell, K.D., Fried, B. (Eds.), World Class Parasites, Food-borne Parasitic Zoonoses - Fish and Plant-Borne Parasites, vol. 11. Springer, New York (USA), pp. 293–334.
Miyamoto, K., Kirinoki, M., Matsuda, H., Hayashi, N., Chigusa, Y., Sinuon, M., et al., 2014. Field survey focused on *Opisthorchis viverrini* infection in five provinces of Cambodia. Parasitol. Int. 63, 366–373.
Montresor, A., Cong, D.T., Sinuon, M., Tsuyuoka, R., Chanthavisouk, C., Strandgaard, H., et al., 2008. Large-scale preventive chemotherapy for the control of helminth infection in Western Pacific countries: six years later. PLoS Neglect. Trop. Dis. 2, e278.
Muth, S., 2006. Schistosomiasis mekongi control project towards elimination in Cambodia. Am. J. Trop. Med. Hyg. 75, 182.
Muth, S., Anantaphruti, M.T., Socheat, D., 2003. Intestinal helminthic infections in schoolchildren in Cambodia. Southeast Asian J. Trop. Med. Publ. Health 34, 254–258.
Muth, S., Sayasone, S., Odermatt-Biays, S., Phompida, S., Duong, S., Odermatt, P., 2010. *Schistosoma mekongi* in Cambodia and Lao people's democratic Republic. Adv. Parasitol. 72, 179–203.
Muth, S., Tsuyuoka, R., Socheat, D., Odermatt, P., Ohmae, H., Matsuda, H., et al., 2007. Control of *Schistosoma mekongi* in Cambodia: results of eight years of control activities in the two endemic provinces. Trans. R. Soc. Trop. Med. Hyg. 101, 34–39.
National Centre for Parasitology Entomology and Malaria Control, 2011. Control of Neglected Tropical Diseases in Cambodia. An Integrated National Plan of Action Focused on Diseases Controlled and Eliminated by Preventive Chemotherapy, 2011–2015. National Centre for Parasitology Entomology and Malaria Control, Phnom Penh (Cambodia).
Ngoenkla, R., Piangjai, S., Somwang, P., Moophayak, K., et al., 2010. Emerging helminths infections in snails and cyprinoid fish in sewage treatment wetland waters in Cambodia. Asian J. Water Environ. Pollut. 7, 13–21.
Olveda, R., Leonardo, L., Zheng, F., Sripa, B., Bergquist, R., Zhou, X.-N., 2010. Coordinating research on neglected parasitic diseases in Southeast Asia through networking. Adv. Parasitol. 72, 55–77.
Parish, R.A., 1985. Intestinal parasites in Southeast Asian refugee children. West. J. Med. 143, 47–49.
Park, S.K., Kim, D.-H., Deung, Y.-K., Kim, H.-J., Yang, E.-J., Lim, S.-J., et al., 2004. Status of intestinal parasite infections among children in Bat Dambang, Cambodia. Kor. J. Parasitol. 42, 201–203.

Petney, T., Sithithaworn, P., Andrews, R., Kiatsopit, N., Tesana, S., Grundy-Warr, C., et al., 2012. The ecology of the *Bithynia* first intermediate hosts of *Opisthorchis viverrini*. Parasitol. Int. 61, 38–45.

Petney, T.N., 2001. Environmental, cultural and social changes and their influence on parasite infections. Int. J. Parasitol. 31, 919–932.

Petney, T.N., Andrews, R.H., Saijuntha, W., Wenz-Mücke, A., Sithithaworn, P., 2013. The zoonotic, fish-borne liver flukes *Clonorchis sinensis, Opisthorchis felineus* and *Opisthorchis viverrini*. Int. J. Parasitol. 43, 1031–1046.

Piraksakulrat, O., Sithithaworn, P., Laoporn, N., Laha, T., Petney, T.N., et al., 2013. A crosssectional study on the potential transmission of the carcinogenic liver fluke *Opisthorchis viverrini* and other fishborne zoontic trematodes by aquiculture fish. Foodb. Pathog. Dis. 10, 35–41.

Rim, H.J., 2005. Clonorchiasis: an update. J. Helminthol. 79, 269–281.

Schär, F., Inpankaew, T., Traub, R.J., Khieu, V., Dalsgaard, A., Chimnoi, W., et al., 2014. The prevalence and diversity of intestinal parasitic infections in humans and domestic animals in a rural Cambodian village. Parasitol. Int. 63, 597–603.

Segal, D.B., Humphery, J.M., Edwards, S.J., Kirby, M.D., 1968. Parasites of man and domestic animals in Vietnam, Thailand, Laos, and Cambodia: host list and bibliography. Exp. Parasitol. 23, 412–464.

Selstad Utaaker, K., Robertson, L.J., 2015. Climate change and foodborne transmission of parasites: a consideration of possible interactions and impacts for selected parasites. Food Res. Int. 68, 16–23.

Sithithaworn, P., Andrews, R., Shekhovtsov, S.V., Mordvinov, V.A., Furman, D.P., 2014. Helminth - trematode: *Opisthorchis viverrini* and *Opisthorchis felineus*. In: Motarjemi, Y. (Ed.), Encyclopedia of Food Safety. Academic Press, Waltham (USA), pp. 170–178.

Sithithaworn, P., Andrews, R.H., Nguyen Van, D., Wongsaroj, T., Sinuon, M., Odermatt, P., et al., 2012a. The current status of opisthorchiasis and clonorchiasis in the Mekong Basin. Parasitol. Int. 61, 10–16.

Sithithaworn, P., Yongvanit, P., Tesana, S., Pairojkul, C., 2007. Liver flukes. In: Murrell, K.D., Fried, B. (Eds.), World Class Parasites, Food-borne Parasitic Zoonoses - Fish and Plant-Borne Parasites, vol. 11. Springer, New York (USA), pp. 3–52.

Sithithaworn, P., Ziegler, A.D., Grundy-Warr, C., Andrews, R.H., Petney, T.N., 2012b. Alterations to the lifecycle of liverflukes: dams, roads, and ponds. Lancet Infect. Dis. 12, 588.

Sohn, W.-M., Kim, H.-J., Yong, T.-S., Eom, K.S., Jeong, H.-G., Kim, J.-K., et al., 2011a. *Echinostoma ilocanum* infection in Oddar Meanchey province, Cambodia. Kor. J. Parasitol. 49, 187–190.

Sohn, W.-M., Shin, E.-H., Yong, T.-S., Eom, K.S., Jeong, H.-G., Sinuon, M., et al., 2011b. Adult *Opisthorchis viverrini* flukes in humans, Takeo, Cambodia. Emerg. Infect. Dis. 17, 1302–1304.

Sohn, W.M., Chai, J.Y., Yong, T.S., Eom, K.S., Yoon, C.H., Sinuon, M., et al., 2011c. *Echinostoma revolutum* infection in children, Pursat province, Cambodia. Emerg. Infect. Dis. 17, 117–119.

Sohn, W.M., Yong, T.S., Eom, K.S., Pyo, K.H., Lee, M.Y., Lim, H., et al., 2012. Prevalence of *Opisthorchis viverrini* infection in humans and fish in Kratie province, Cambodia. Acta Trop. 124, 215–220.

Sripa, B., Brindley, P.J., Mulvenna, J., Laha, T., Smout, M.J., Mairiang, E., et al., 2012. The tumorigenic liver fluke *Opisthorchis viverrini* – multiple pathways to cancer. Trends Parasitol. 28, 395–407.

Sripa, B., Kaewkes, S., Intapan, P.M., Maleewong, W., Brindley, P.J., 2010. Food-borne trematodiases in Southeast Asia: epidemiology, pathology, clinical manifestation and control. Adv. Parasitol. 72, 305–350.

Stauffer, W.M., Sellman, J.S., Walker, P.E., 2004. Biliary liver flukes (opisthorchiasis and clonorchiasis) in immigrants in the United States: often subtle and diagnosed years after arrival. J. Trav. Med. 11, 157–160.

Stauffer, W.M., Sellman, J.S., Walker, P.F., 2003. Liver flukes (opistorchiasis and clonorchiasis) in immigrants in the United States: often subtle and diagnosed years after arrival. Am. J. Trop. Med. Hyg. 69, 257–258.

Stich, A.H.R., Biays, S., Odermatt, P., Men, C., Saem, C., Sokha, K., et al., 1999. Foci of *Schistosomiasis mekongi*, northern Cambodia: II. Distribution of infection and morbidity. Trop. Med. Int. Health 4, 674–685.

Story, M., Harris, L.J., 1989. Food habits and dietary change of Southeast Asian refugee families living in the United States. J. Am. Diet Assoc. 89, 800–803.

Tateno, S., Takeuchi, T., Kobayashi, S., Tanabe, M., Mirua, S., Asami, K., 1981. Health status of indochina refugees in Japan: prevalence of intestinal parasites and eosinophilia. Jpn. J. Parasitol. 30, 485–492.

Thaenkham, U., Nuamtanong, S., Sa-nguankiat, S., Yoonuan, T., Touch, S., Manivong, K., et al., 2010. Monophyly of *Opisthorchis viverrini* populations in the lower Mekong Basin, using mitochondrial DNA nad1 gene as the marker. Parasitol. Int. 59, 242–247.

Thaenkham, U., Nuamtanong, S., Vonghachak, Y., Yoonuan, T., Sanguankiat, S., Dekumyoy, P., et al., 2011. Discovery of *Opisthorchis lobatus* (trematoda: *Opisthorchiidae*): a new record of small liver flukes in the Greater Mekong Sub-region. J. Parasitol. 97, 1152–1158.

Torgerson, P.R., Devleesschauwer, B., Praet, N., Speybroeck, N., Willingham, A.L., Kasuga, F., et al., 2015. World Health Organization estimates of the global and regional disease burden of 11 foodborne parasitic diseases, 2010: a data synthesis. PLoS Med. 12, e1001920.

Touch, S., Komalamisra, C., Radomyos, P., Waikagul, J., 2009. Discovery of *Opisthorchis viverrini* metacercariae in freshwater fish in southern Cambodia. Acta Trop. 111, 108–113.

Touch, S., Yoonnan, T., Nuamtanong, S., Homsuwan, N., Phuphisut, O., Thaenkham, U., et al., 2013. Seasonal variation of *Opisthorchis viverrini* metacercarial infection in cyprinid fish from southern Cambodia. Trop. Med. Parasitol. 36, 1–7.

Traub, R.J., Macaranas, J., Mungthin, M., Leelayoova, S., Cribb, T., Murrell, K.D., et al., 2009. A new PCR-based approach indicates the range of *Clonorchis sinensis* now extends to central Thailand. PLoS Neglected Trop. Dis. 3, e367.

United Nations Population Division, 2017. World Population Prospects: The 2017 Revision. Retrieved from: https://population.un.org/wpp/Download/Standard/Population/.

Vernes, A., Abdellatifi, M., Deicas, E., Poirriez, J., Abdelmalek, R.Y., Leroux, N., et al., 1982. Parasitic infestation and hemoglobinopathies in a population of 1170 refugees from Southeast Asia. Nouv. Presse Med. 11, 2687–2691.

Vonghachack, Y., Odermatt, P., Taisayyavong, K., Phounsavath, S., Akkhavong, K., Sayasone, S., 2017. Transmission of *Opisthorchis viverrini, Schistosoma mekongi* and soil-transmitted helminthes on the Mekong islands, southern Lao PDR. Infect. Dis. Poverty 6, 131.

Wongsawad, C., Wongsawad, P., 2012. *Opisthorchis viverrini* and *Haplorchis taichui*: development of a multiplex PCR assay for their detection and differentiation using specific primers derived from HAT-RAPD. Exp. Parasitol. 132, 237–242.

World Health Organization, 1995. Control of Foodborne Trematode Infections : Report of a WHO Study Group. WHO, Geneva (Switzerland).

World Health Organization, 2008. Review on the Epidemiological Profile of Helminthiases and Their Control in the Western Pacific Region, 1997–2008. WHO WPRO, Manila (Philippines).

World Health Organization, 2009. First Mekong-Plus Programme Managers Workshop on Lymphatic Filariasis and Other Helminthiasis. Phnom Penh, Cambodia, 23–26 March 2009. WHO WPRO, Manila (Philippines).

World Health Organization, 2011. Report of the WHO Expert Consultation on Foodborne Trematode Infections and Taeniasis/cysticercosis. Vientiane, Lao People's Democratic Republic 12–16 October 2009. WHO, Geneva (Switzerland).

World Health Organization, 2015. Schistosomiasis: number of people treated worldwide in 2013. Wkly. Epidemiol. Rec. 90, 25–32.

World Health Organization, Food and Agriculture Organization, 2004. Report of the Joint WHO/FAO Workshop on Foodborne Trematode Infections in Asia. WHO, Manila (Philippines).

Xayaseng, V., Phongluxa, K., van Eeuwijk, P., Akkhavong, K., Odermatt, P., 2013. Raw fish consumption in liver fluke endemic areas in rural southern Laos. Acta Trop. 127, 105–111.

Yang, G.-J., Utzinger, J., Lv, S., Qian, Y.-J., Li, S.-Z., Wang, Q., et al., 2010. The regional Network for Asian schistosomiasis and other helminth zoonoses (RNAS+): target diseases in face of climate change. Adv. Parasitol. 73, 101–135.

Yangco, B.G., De Lerma, C., Lyman, G.H., Price, D.L., 1987. Clinical study evaluating efficacy of praziquantel in clonorchiasis. Antimicrob. Agents Chemother. 31, 135–138.

Yong, T.-S., Chai, J.-Y., Sohn, W.-M., Eom, K.S., Jeoung, H.-G., Hoang, E.-H., et al., 2014. Prevalence of intestinal helminths among inhabitants of Cambodia (2006–2011). Kor. J. Parasitol. 52, 661–666.

Yong, T.S., Shin, E.H., Chai, J.Y., Sohn, W.M., Eom, K.S., Lee, D.M., et al., 2012. High prevalence of *Opisthorchis viverrini* infection in a riparian population in Takeo province, Cambodia. Kor. J. Parasitol. 50, 173–176.

Ziegler, A.D., Petney, T.N., Grundy-Warr, C., Andrews, R.H., Baird, I.G., Wasson, R.J., et al., 2013. Dams and disease triggers on the lower Mekong River. PLoS Neglected Trop. Dis. 7, e2166.

CHAPTER FOUR

Challenges and Lessons From Conducting A Paediatric Clinical Trial in Sub-Saharan Africa: The Case of the Praziquantel Oral Dispersible Tablets Phase II Study in Côte d'Ivoire

Eliézer N'Goran*, N'da Angbeletchi David Aka*, Mamadou Ouattara*, Eric Huber[¶,||], Deon Bezuidenhout[#] and Elly Kourany-Lefoll[1,§], on behalf of the Pediatric Praziquantel Consortium

*Université Félix Houphouët-Boigny, Abidjan, Côte d'Ivoire
[§]Ares Trading S.A., Switzerland, an affiliate of Merck KGaA, Darmstadt, Germany
[¶]Swiss Tropical and Public Health Institute, Basel, Switzerland
[||]University of Basel, Basel, Switzerland
[#]Merck (Pty) Ltd, South Africa, an affiliate of Merck KGaA, Darmstadt, Germany
[1]Corresponding author: E-mail: elly.kourany-lefoll@merckgroup.com

Contents

Abstract

The importance of implementing paediatric clinical trials for neglected tropical diseases (NTDs) in compliance with the Good Clinical Practices of the International Conference

Advances in Parasitology, Volume 103
ISSN 0065-308X
https://doi.org/10.1016/bs.apar.2018.09.002

of Harmonisation (ICH-GCP) and other applicable regulatory and ethics guidelines is increasingly being recognised as an essential pathway to provide safe and effective medicines for millions of untreated children living in sub-Saharan Africa (SSA). This paper describes the learnings and challenges faced by the Pediatric Praziquantel Consortium team during the implementation of an industry-sponsored Phase II clinical study in pre-school-aged children infected with schistosomiasis, conducted in remote rural settings in Côte d'Ivoire. The importance of close interactions with the ethics committee, the regulatory and administrative authorities and the rural communities are highlighted. The difficulties faced included obtaining a valid informed consent from the child's parent or guardian, the collection of blood samples from children during the study while respecting cultural beliefs as well as the weak medical research infrastructure. The paper illustrates how a public-private collaborative partnership can promote capacity-building and high-quality NTD paediatric clinical research in SSA.

1. INTRODUCTION

At the General Assembly of the United Nations in September 2015, countries committed to the Sustainable Development Goals including Goal 3: "Ensure healthy lives and promote well-being for all at all ages." One target of this goal is to reduce, by 2030, the under-5-year-old mortality rate to at least as low as 25 per 1000 live births (Anon, 2015). It is known that the highest rate of childhood deaths concentrates primarily in low- and middle-income countries (LMICs), mainly in sub-Saharan Africa (SSA), where one child in 12 dies before the age of 5 years, compared with one in 147 in high-income countries (Molyneux and Molyneux, 2016).

Paediatric clinical trials in SSA are an essential pathway to provide safe and effective medicines to children living in these countries. Many of these children live in remote areas in extreme poverty and carry chronic and debilitating parasitic infections such as intestinal helminth infections — ascariasis, trichuriasis, hookworm infection and intestinal schistosomiasis, or other neglected tropical diseases (NTDs) such as urogenital schistosomiasis, trachoma, leprosy and Buruli ulcer. These children are often not treated due to the lack of appropriate paediatric medication. As highlighted by the World Health Organization (WHO), there is insufficient evidence to support the use of current dosages and formulations of medicines included on the WHO Essential Medicine List for the treatment of NTDs in the paediatric population (Al-Saffer and Ridge, 2010). Recent legislation and regulatory initiatives, including the United States Best Pharmaceuticals for Children Act (BCPA) and the Pediatric Research Equity Act (PREA), as well as the Paediatric Regulation in Europe (Anon, 2002, 2006, 2007) required the pharmaceutical industry to conduct high-quality research in the development of medicines for children to ensure that they are specifically authorised for that use, with appropriate forms and formulations and to guarantee the availability of high quality information about treatments used by children. High-quality and ethical medical research to identify treatments for NTDs in this vulnerable population of children is therefore important and necessary.

In order to ensure equity and maintain high standards of quality in paediatric NTD medical research carried out in SSA countries, several factors have to be considered including:

- The capacity of the local ethics committees and National Medical Regulatory Authority (NMRA) to oversee the medical research in order to ensure the protection of the participants and the quality and scientific integrity of the clinical data. Although there is progressive improvement in regulatory capacity in many SSA countries, WHO estimates that at least 30% of NMRAs in Africa have limited capacity to perform core regulatory functions (Ndomondo-Sigonda et al., 2017).
- The scarce availability of the required research infrastructure, equipment and qualified human resources with the necessary skills, competencies and experience in clinical research practice.
- The low levels of education and revenues in the affected communities and populations, which may influence the understanding of the research concepts discussed and/or compromise the motivation of the participants to enrol in the trial.
- The weak civil registration and vital statistics (CRVS) systems, as well as various deleterious cultural or socio-political contexts. According to the United Nations Children's Emergency Fund (UNICEF) global databases (2013) (Anon, 2013a), only 44% of children are registered by their fifth birthday in sub-Saharan Africa, leaving about 85 million children under the age of five unrecorded. Despite the well-documented benefits of CRVS, including its importance in the health sector in relation to child health and disease-specific mortality, many countries in SSA do not have adequate systems in place.

Schistosomiasis, also called Bilharzia, is an important tropical disease caused by parasitic worms which have infectious larvae typically found in environmental freshwater bodies. It is one of the most important NTDs, killing an estimated 200,000 people each year in the African region alone. Estimates show that at least 4 million people required preventive treatment for schistosomiasis in 2016 (Anon, 2018c). Schistosomiasis is prevalent in tropical and sub-tropical areas, especially in poor communities without access to safe drinking water and adequate sanitation. It is estimated that at least 91.4% of those requiring treatment for schistosomiasis live in SSA, particularly in rural, agricultural and peri-urban areas. In many areas, schistosomiasis infects a large proportion of children under the age of 14 years. The WHO-recommended strategy to control morbidity caused by schistosomiasis is based on preventive chemotherapy interventions using a single oral dose (40 mg/kg) of praziquantel (PZQ) tablets, targeting the majority of the at-risk population. Pre-school-aged children (PSAC), who are a high-risk group for schistosome infections, are currently not included in the WHO schistosomiasis control programmes, due to the lack of a suitable paediatric formulation of PZQ.

The Pediatric Praziquantel Consortium (Anon, 2018b) is a public-private partnership formed in July 2012 with the goal of developing an oral dispersible tablet (ODT) formulation of PZQ for treatment of PSAC with schistosomiasis and obtaining regulatory approval of this formulation by a stringent regulatory authority in order to facilitate local endemic country approvals, WHO pre-qualification and future access. The Consortium supports the WHO strategic plan for eliminating soil-transmitted helminths and schistosomiasis as a public health problem in children, which aims to increase global PSAC deworming coverage to 50% by 2015 and 75% by 2020.

Two paediatric clinical trials in SSA have been conducted to date in the frame of the Consortium's PZQ paediatric formulation development programme—a taste study in African children aged 6 to 11 years (NCT02315352) conducted in Tanzania, and Part 1 of a Phase II dose-finding study (NCT02806232) in *Schistosoma mansoni*-infected children aged 2 to 6 years, conducted at the Regional Hospital Centre of Man (Centre Hospitalier Regional de Man; CHRM), in the Tonkpi Region, Côte d'Ivoire, in 2016–17. Additional paediatric studies in schistosome-infected PSAC are planned. This paper describes the experience of the authors in the establishment of a new clinical research facility in the CHRM and illustrates the challenges and opportunities faced during the preparation and implementation of the trial.

2. TRIAL STRUCTURE AND GOVERNANCE

The trial, sponsored by Merck KGaA (Darmstadt, Germany) and managed by the Medicine Department at the Swiss Tropical and Public

Health Institute (Basel, Switzerland) was overseen by the Pediatric Praziquantel Consortium project team. It was mostly funded by a grant received from the Global Health Innovation Technology Fund (GHIT; Tokyo, Japan) and supported by in-kind contributions from the Consortium partners. University Félix Houphouët-Boigny (UFHB), located in Abidjan, provided the investigators responsible for implementing the trial. The study enrolled a total of 420 *S. mansoni*-infected children aged 2 to 6 years, who were randomised to receive doses of Biltricide®, racemate-PZQ ODTs, or laevorotatory-PZQ ODTs.

3. STUDY SETTINGS

Côte d'Ivoire, a French-speaking country in West Africa, has gone through more than a decade of socio-political crises that have degraded the health system and deteriorated the measures for the control of schistosomiasis and other NTDs. The northern and western parts of the country were most affected by these events. In 2010, a 5-year integrated operational plan was developed for six NTD control programmes, assisted by experts from the WHO. Mass drug administration (MDA) control programmes for schistosomiasis, supported by the Schistosomiasis Control Initiative (Imperial College London, United Kingdom) started in 2013 in the endemic regions with an initial emphasis on the Cavally, Guemon and Tonkpi regions in western Côte d'Ivoire, and on the Agnéby region in south-eastern Côte d'Ivoire. At the time of this trial, apart from MDA treatment of school age children, PZQ was hardly available within the Montagne District.

The City of Man, the Capital of the Tonkpi region part of the Montagne District, is located in the western part of the country, about 580 km (approximate 8-hour drive) away from the economic capital, Abidjan, where the trial field coordination office was located. Man has a population of about 200,000 inhabitants, mainly consisting of the Yacoubas, Wobés and Touras ethnic groups. Man was selected because of the large rural catchment area where *S. mansoni* is endemic (Assare et al., 2016) and is representative of cities in which treatment might be expanded in a real-world setting. The recruited participants originated from villages located in the departments of Biankouman and Man in the Tonkpi region, as well as Facobly and Kouibly in the Guémon region (Fig. 1), to minimise the distance between the villages of the participants and the CHRM hospital. Reaching some

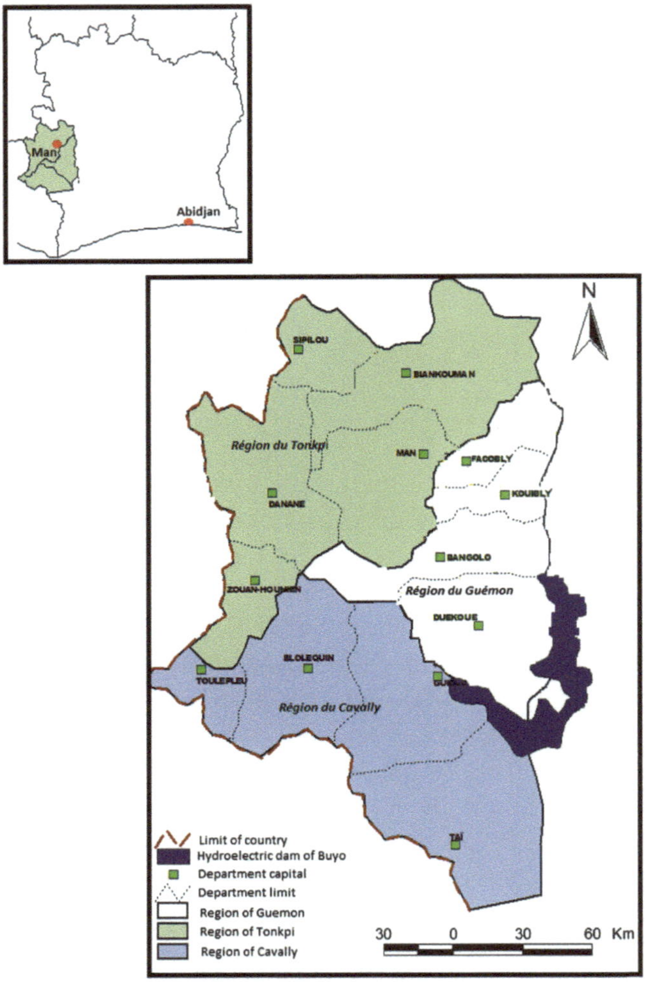

Figure 1 Map of the district Mountain showing the different departments.

villages by a 4-wheel-drive vehicle could take up to 2 h in the dry season, and longer in the rainy season.

The clinical research facility at the CHRM hospital in Man was established by the Consortium clinical trial team, which included researchers from the UFHB. In addition, the Phase II paediatric PZQ study was the first industry-sponsored registration-enabling trial conducted by the investigators and the local research team according to Good Clinical Practices (GCPs) codes of the International Conference of Harmonisation (ICH E6).

4. IMPORTANCE OF COMMUNITY INVOLVEMENT

In order to facilitate trust-building and acceptance of the study, the local team, led by the Investigator, visited Man and Biankouman to discuss the proposed clinical trial with various stakeholders, including the Man Regional Director of Health, the Man and Biankouman Department Directors of Health, the Prefect of the Man region and the Mayor of the city of Man. Information about the trial was subsequently transmitted by the prefect to the sub-prefects and village chiefs. Due to the recurrent leadership crises and erosion of traditional powers at the rural community level, traditional chiefs convened a meeting with their community leaders, in order to discuss and share information about the trial and obtain their acceptance to engage their community members in the study.

After this step, it was necessary to inform and involve the community health staff at the community health centres, the individual village elders and the community as a whole in the setting of community information meetings. Meetings were held in French and with the help of translators, the information was also presented in individual village dialects. The project was executed only after authorisation was granted from the National Ethics Committee (Comité National d'Ethique et de Recherche, CNER) and the administrative and political authorities, including the village authorities. Thereafter, the local study site staff continuously hosted community information days to keep information flow to the communities active, maintain dialogue with community members, and encourage active participation in the trial. This proactive approach proved, as in previous studies (Coulibaly et al., 2017, 2018), to be essential and underscored the success of our recruitment and study conduct.

5. ETHICAL AND REGULATORY APPROVALS

In line with the recommendations from the Council for International Organizations of Medical Sciences (CIOMSs) guidelines and the Nuffield Council on Bioethics, the trial documentation was submitted to Ethics Committee Northwestern/Central Switzerland (EKNZ) and to the Ethics Committee in the host country, the CNER, for ethical clearance. The Principal Investigator was invited to a meeting with CNER to present and discuss the study protocol. The ethical review was completed within two months. Of note, the legal age of parental consent was clarified by CNER as being 21 years, according to the written law (Minority Act, 1970, Article 1), as opposed to 18 years as it is in many other countries. The concept of emancipated minors versus the legal age of majority was not clearly documented in local law and hence required ethical consideration as well. Although Act No 70–483 du 3 AOUT 1970 sur la Minorité considered all married minors as fully emancipated (Article 116), emancipated parents could not provide consent for their children, in contrast to the situation in some other countries in SSA. Children's oral assent as of the age of three years was also requested, though the WHO Ethics Review Committee guidelines state that "researchers should consider asking for assent from children over the age of seven years" (Anon).

Access to care was carefully discussed and included into the protocol: children infected with *S. mansoni* who could not be enrolled in the trial, as well as children who were not cured after PZQ treatment, were able to receive treatment with PZQ tablets (Cesol® 600 mg) at a dose of 40 mg/kg. In addition, two members from CNER made a field visit to the study site to monitor trial progress and participants safety. Close interactions with the CNER was therefore deemed essential to comply with local ethical requirements.

However, it was unclear whether a formal regulatory clinical trial application (CTA) submission was needed to obtain official authorisation for the trial by the NMRA, as information about clinical trial regulations was not published on the Ministry of Health website. At the time of trial start in 2016, the Ministry of Health was in the process of constituting its regulatory health authority and CTA requirements, including the related pharmacovigilance reporting, which did not have an established process. The Sponsor and trial investigators held constructive meetings with the Direction de la Pharmacie du Médicament et des Laboratoires (DPMLs) and other Ministry of Health representatives, to discuss the study protocol and process for obtaining the trial authorisations needed. The outcome of these continued interactions encouraged the DMPL to set up, in January 2018, an external evaluation process in the frame of CTA dossiers (Anon, 2018a). This experience underscores the necessity for the Sponsor and stakeholders to utilise the local knowledge of trial investigators and continuous dialogue with regulators to navigate an unclear regulatory environment.

6. CAPACITY-BUILDING

One challenge that is generally faced by study teams conducting clinical trials on NTDs in LMICs is the capacity of the clinical trial site to conduct the study. The treatment facilities, often located in remote rural areas, are usually small, with limited human resources and often basic infrastructure. The CIOMS guideline (Anon, 2016a) states: "As part of their obligation, sponsors, and researchers must make every effort, to assist in building local research capacity. In some cases, in order to ensure an overall fair distribution of the benefits and burdens of the research, additional benefits such as investments in the local health infrastructure should be provided to the population or community". However, in general, Sponsor companies favour investigational sites that already have infrastructure, including facilities and equipment, as well as past experience in place to enable them to conduct a clinical trial.

To perform the study at the CHRM, the trial team set out to build a dedicated clinical trial research facility on the hospital campus, by applying major upgrades to an old hospital storeroom. The new building was located in proximity to the hospital pharmacy, hospital laboratory and emergency department. The facility was air-conditioned and consisted of a 12-bed clinic room with an attached medical examination room, office space, two showers and two toilets. Local knowledge and skills were used to do the actual construction work, which was the first step into providing local employment to a small team of contractors. The main challenge during the construction was to ensure the process stayed on track to meet the required timelines. Later, an office building and boardroom were added to the existing clinic ward, to further enhance site capacity.

Development of the trial site staff was another major activity. The study staff consisted of 27 people. Due to difficulties in finding qualified research personnel, the team negotiated with the hospital to allocate one qualified paediatrician, two medical doctors, two pharmacists, two nurses, one biologist and three laboratory technicians to support the study on a part-time basis. The rest of the study team (the PI, four co-investigators, a laboratory technician, a paediatrician, a medical doctor, two nurses, four nurse aids and two computer scientists) were from the UFHB (or sub-contracted by the UFHB). The recruited local healthcare personnel as well as several staff members from the UFHB had not previously conducted clinical trials, hence a sustained and continuous effort was needed to train them on ICH-GCP requirements, with repeated explanations on medical research techniques such as appropriate documentation, randomisation, safety reporting, etc. The skills shortage was also addressed by extended site initiation visits, and more frequent and longer monitoring visits that included additional training.

7. INFORMED CONSENT AND ORAL ASSENT

A standardised instruction guide for obtaining the informed consent was written explaining the steps to be taken for organising community meetings in the villages, the process of individual information to be communicated in the households and the signature process of the consent form. The involvement of fieldworkers with knowledge of local population and languages was important in facilitating recruitment of study participants.

A practical challenge the team faced was obtaining paediatric consent and assent. The International GCP code (ICH-GCP E6 and WHO-GCP) requires that when a minor is enrolled in a clinical trial, informed consent is given by the parent or a legally authorised representative (LAR). However, the definition of LAR was difficult to apply in rural study settings, since the guardians of children were usually relatives or community members, without a formal nomination by a tribunal. During the site audit conducted by the Sponsor, one of the audit findings was the inadequacy of the informed consent process due to the lack of documented evidence at the site to confirm the legal representatives/guardians' identity and their legal authorisation to act on behalf of the child.

The lack of availability of medical files and legal documents such as birth certificates and identification documents did result in challenging processes to adequately identify study participants and their parents/guardians. In the demographic setting of the study, medical records are rarely available, as the study population has very limited access to medical treatment, many children are born at home, and birth certificates may not be issued. In Côte d'Ivoire, the birth registration coverage is estimated to be around 55% (Anon, 2013b). Patient records may only be available if the child has participated in a vaccination programme and/or has been admitted to the local hospital in recent weeks.

In our study, several children had to be excluded, as the investigator was required to verify the identity and age of the participant by provision of a birth certificate or certificate of vaccination, and these documents were often unavailable. In case identification documents of the parent/guardian were unavailable, the village chief was requested to confirm their identity. The response was noted in the source documents.

In addition, most of the parents/guardians were unable to write. Those unable to write provided their thumbprints and an independent witness signed the informed consent form (ICF) to confirm that the written information had been adequately explained to the parents/guardians and that all questions had been answered to the participant's satisfaction. In the geographic location of the study, it was sometimes difficult to find independent witnesses to co-sign the ICF, as most of the population was illiterate. The same witness was therefore used for several participants.

Oral assent from children above three years of age was also obtained. Each child above three years of age had the opportunity to state verbally if he or she would consent to the study and this was documented in the ICF on the signature page of the Certificate of Consent. In case of refusal, the child was not enrolled in the study even if his parent or guardian had provided consent. Training was performed for the staff involved in the ICF process to improve the quality of the documentation and to enable them to ensure that the ICF is correctly completed.

8. BLOOD COLLECTION

The importance of continuous community sensitisation was experienced by the team, when recruitment came to a virtual standstill. The team discovered that by having one parent voicing concern about the collection of blood samples (by venous sampling and finger pricking) needed for the safety laboratory tests and pharmacokinetics (PKs) analyses, the whole community shied away from participation into the trial. As reported by several authors (Grietens et al., 2014; O'Neill et al., 2016), collecting blood and other biological samples from individuals recruited in clinical trials in SSA can be challenging, mainly due to the local rumours surrounding "blood stealing" or "blood selling," associated with conceptions of blood as a life force. Such fears can potentially have direct effects on the success of research projects. Our study was unable to enrol patients for several weeks, due to this concern. To reactivate recruitment, several community mobilisation meetings at the local villages were held, where the meaning and reasons for collecting blood samples was explained, including the whole procedure of finger pricking, capillary blood drawing, dried-blood spot testing as well as transporting the blood samples to Germany for bioanalytical sample analyses. Maintaining recruitment meant continuous dialogue with the communities, which underscores the value of close collaboration on a continuous basis. In this respect regulatory agencies, that require PK/pharmacodynamic studies to be generated in one or several paediatric age ranges as part of the paediatric medicines development extrapolation plan (Anon, 2016b), could be made more aware of the challenges in PK blood sampling in children living in SSA while at the same time demonstrating respect for community values.

9. OTHER CHALLENGES

In general, pharmaceutical industry sponsors require trial sites to be able to complete the case report forms in English. This requirement did place a challenge on our local study team. In Côte d'Ivoire, there is a shortage of well-trained, fluent English-speaking medical staff. In the rural settings of Man, English is not used and very little of it is spoken in the area. Having a language barrier affected the transfer of the data collected in French to the English database, and resulted in higher query rates and slower query resolution. The team relied very much on the fluent English-speaking monitoring team to assist the site in explaining queries, in order for the site staff to understand what was queried. The language barrier was also overcome by using a creative electronic data collection platform, with drop-down data fields in French to describe the data needed to be entered, and by using appropriate medical online dictionaries for the translation of medical terms.

Another issue faced was the political instability during the trial conduct. A major mutiny broke out among the army of Côte d'Ivoire in January 2017, which had spread to nine cities, including Man. A second mutiny broke out among the same soldiers in May 2017. This political turmoil caused significant delays in the implementation of the study. The staff could not go to work for security reasons as this would entail a risk to their life.

10. DISCUSSION

Published literature (Joseph et al., 2016; Mbuagbaw et al., 2011; Molyneux and Molyneux, 2016) highlight the complexities and difficulties of conducting paediatric trials in LMICs, due to economic scarcities, cultural beliefs, and local customs. Our experience in Côte d'Ivoire adds further evidence to these published data and supports the recommendations from Ravinetto et al. (2016) to revise the international GCP guidelines and to consider the challenges of clinical research in low-resource settings. In our case, the informed consent procedure for minors and dealing with illiterate parents/guardians was a key challenge. Blood sampling in children was another major challenge and our recommendation is that international guidelines on PK extrapolation in paediatric development consider the difficulties of PK sampling in view of the socio-cultural beliefs in these countries.

Our study was a good example of how a public-private collaborative partnership, including partners from high-income countries, can promote capacity-building and high-quality NTD paediatric clinical research in a SSA country that has suffered from civil and political unrest (Bonfoh et al., 2011). Despite the difficulties and hurdles faced, a new clinical trial facility was established in the CHRM that is able to implement industry- and investigator-sponsored paediatric trials for NTDs in compliance with ICH-GCP and international ethics guidelines. The competence of the clinical research site and the study staff increased over the course of the study. The challenges and pragmatic mitigation strategies highlighted for research problems at all levels in this first industry sponsored study conducted in the CHRM are of value to other partners and actors that may be considering rolling out trials on NTDs affecting paediatric populations in SSA.

One aspect that is worth highlighting is the sustainability of the research facility in the CHRM to conduct future research. This will require continuous involvement in the conduct of phase II/III clinical studies on NTD drugs under development. The Consortium is planning to conduct a phase III trial in shistosomiasis infected children in the CHRM. The partnership established between the UFHB and the CHRM may become the frame for additional clinical trial research collaborations. The capacity of the site to conduct clinical research will also be promoted.

Paediatric research infrastructure to conduct NTD studies in SSA is a growing need and although some partnerships, such as the African Malaria Network Trust (AMANET) and the Gates Malaria Partnership (GMP) are examples of how capacity strengthening can promote high-quality research in SSA, these networks do not specifically address the paediatric population affected with NTDs. In addition, the current paediatric clinical trial infrastructure in these countries remains fragmented, not well resourced, unorganised, and inefficient. Recently, the Innovative Medicines Initiative issued a call (IMI2-2016–10–04) for the creation of a pan-European paediatric clinical trials network that will facilitate the development and availability of new drugs and other therapies and the expansion of knowledge about drugs currently in practice for the entire paediatric population in Europe. A similar multidisciplinary public-private consortium collaborative effort is required that will provide a ready and transparent mechanism for assembling key stakeholders to develop an integrative, open and sustainable paediatric clinical trial network in SSA, in order to meet the challenges described and be transformative to address collectively neglected children's unmet needs for essential medicines.

ACKNOWLEDGEMENTS

In addition to contributions by all partners, the Pediatric Praziquantel Consortium receives funding from Merck KGaA, Darmstadt, Germany, and grant support from the Bill & Melinda Gates Foundation (BMGF; Grant no. OPP1063223), and the Global Health Innovative Technology (GHIT) Fund (Grant nos. 2013–212, 2014–206 and 2016–110). Furthermore, this project is part of the second European & Developing Countries Clinical Trials Partnership (EDCTP2) programme supported by the European Union (grant number RIA2016S-1641). The BMGF, the GHIT Fund and the EDCTP had no role in the decision to publish or preparation of the manuscript.

Disclosures

EK-L is an employee of Ares Trading S.A., Switzerland, an affiliate of Merck KGaA, Darmstadt, Germany and DB is an employee of Merck (Pty) Ltd., South Africa, an affiliate of Merck KGaA, Darmstadt, Germany. All other authors have declared that no competing interests exist.

REFERENCES

Al-Saffer, R., Ridge, A.L., 2010. Overview of the Evidence for Safety and Efficacy of Medicines for the Treatment of Neglected Tropical Diseases in Children. Geneva.

Anon The Process of Obtaining Informed Consent World Health Organization Research Ethics Review Committee.

Anon, 2002. Best Pharmaceuticals for Children Act, Title IV (Public Law 107-109).

Anon, 2006. Regulation (EC) no 1901/2006 of the European parliament and of the Council of 12 December 2006 on medicinal products for paediatric use and amending the regulation (EEC) No 1768/92, Directive 2001/20/EC, Directive 2001/83/EC and regulation (EC) No 726/2004. Official J Eur Union L378, 1–19.

Anon, 2007. Pediatric Research Equity Act, Title V (Public Law 110-85).

Anon, 2013a. Every Child's Birthright; Inequities and Trends in Birth Registration.

Anon, 2013b. Potential and Principles for Health Sector Actions to Strengthen Civil Registration and Vital Statistics Systems.

Anon, 2015. Transforming Our World: The 2030 Agenda for Sustainable Development.

Anon, 2016a. International Ethical Guidelines for Health-related Research Involving Humans.

Anon, 2016b. Reflection Paper on Extrapolation of Efficacy and Safety in Paediatric Medicine Development.

Anon, 2018a. Modification de la décision n°1132 du 19 avril 2017 instituant des redevances pour les demandes d'Autorisation d'essais cliniques) publiée le 2 Janvier 2018. Direction de la Pharmacie, du Médicament et des Laboratoires de Côte d'Ivoire.

Anon, 2018b. Pediatric Praziquantel Corsortium. Pediatric Praziquantel Corsortium. Available: https://www.pediatricpraziquantelconsortium.org/.

Anon, 2018c. Schistosomiasis.

Assare, R.K., Hurlimann, E., Ouattara, M., N'guessan, N.A., Tian-Bi, Y.N., Yapi, A., Yao, P.K., Coulibaly, J.T., Knopp, S., N'goran, E.K., Utzinger, J., 2016. Sustaining the control of *Schistosoma mansoni* in western Cote d'Ivoire: baseline findings before the implementation of a randomized trial. Am. J. Trop. Med. Hyg. 94, 352–360.

Bonfoh, B., Raso, G., Koné, I., Dao, D., Girardin, O., Cissé, G., Zinsstag, J., Utzinger, J., Tanner, M., 2011. Research in a war zone. Nature 474, 569–571.

Coulibaly, J.T., Panic, G., Silue, K.D., Kovac, J., Hattendorf, J., Keiser, J., 2017. Efficacy and safety of praziquantel in preschool-aged and school-aged children infected with *Schistosoma mansoni*: a randomised controlled, parallel-group, dose-ranging, phase 2 trial. Lancet Glob Health 5, e688–e698.

Coulibaly, J.T., Panic, G., Yapi, R.B., Kovac, J., Barda, B., N'gbesso, Y.K., Hattendorf, J., Keiser, J., 2018. Efficacy and safety of ascending doses of praziquantel against *Schistosoma haematobium* infection in preschool-aged and school-aged children: a single-blind randomised controlled trial. BMC Med. 16, 81.

Grietens, K.P., Ribera, J.M., Erhart, A., Hoibak, S., Ravinetto, R.M., Gryseels, C., Dierickx, S., O'neill, S., Muela, S.H., D'alessandro, U., 2014. Doctors and vampires in sub-Saharan Africa: ethical challenges in clinical trial research. Am. J. Trop. Med. Hyg. 91, 213–215.

Joseph, P.D., Caldwell, P.H.Y., Tong, A., Hanson, C.S., Craig, J.C., 2016. Stakeholder views of clinical trials in low- and middle-income countries: a systematic review. Pediatrics 137 e20151766.

Mbuagbaw, L., Thabane, L., Ongolo-Zogo, P., Lang, T., 2011. The challenges and opportunities of conducting a clinical trial in a low resource setting: the case of the Cameroon mobile phone SMS (CAMPS) trial, an investigator initiated trial. Trials 12, 145.

Molyneux, M., Molyneux, E., 2016. Reaching millennium development goal 4. Lancet Glob Health 4, e146–e147.

Ndomondo-Sigonda, M., Miot, J., Naidoo, S., Dodoo, A., Kaale, E., 2017. Medicines regulation in Africa: current state and opportunities. Pharmaceut. Med. 31, 383–397.

O'neill, S., Dierickx, S., Okebe, J., Dabira, E., Gryseels, C., D'alessandro, U., Peeters Grietens, K., 2016. The importance of blood is infinite: conceptions of blood as life force, rumours and fear of trial participation in a Fulani village in rural Gambia. PLoS One 11 e0160464.

Ravinetto, R., Tinto, H., Diro, E., Okebe, J., Mahendradhata, Y., Rijal, S., Gotuzzo, E., Lutumba, P., Nahum, A., De Nys, K., Casteels, M., Boelaert, M., 2016. It is time to revise the international Good Clinical Practices guidelines: recommendations from non-commercial North–South collaborative trials. BMJ Global Health 1 e000122.

CHAPTER FIVE

Drug Combinations Against Soil-Transmitted Helminth Infections

Wendelin Moser[*,§], **Christian Schindler**[*,§] **and Jennifer Keiser**[*,§,1]
[*]Swiss Tropical and Public Health Institute, Basel, Switzerland
[§]University of Basel, Basel, Switzerland
[1]Corresponding author: E-mail: jennifer.keiser@swisstph.ch

Contents

Abstract

The soil-transmitted helminths (STHs), *Ascaris lumbricoides*, hookworm and *Trichuris trichiura* are common in areas with warm and moist climates with little access to adequate water, sanitation, and hygiene affecting the poorest populations. The current control strategy of the World Health Organization is preventive chemotherapy (PC), i.e., the administration of the two benzimidazoles (albendazole and mebendazole) using single, oral doses to at risk populations without prior diagnosis. The recent success of PC is threatened by anthelmintic drug resistance and the low efficacy of

Advances in Parasitology, Volume 103
ISSN 0065-308X
https://doi.org/10.1016/bs.apar.2018.08.002

the drugs against hookworm (mebendazole) and *T. trichiura* (albendazole and mebendazole). Only a handful of alternative drugs with anthelmintic properties are available, however, none of the drugs show high efficacy against all three STHs. The combination of two drugs with different activity profiles presents an attractive alternative, which could prevent the development of drug resistance and increase the efficacy compared to monotherapy. In this review, we summarize the efficacy of current and alternative anthelmintics, coadministrations and triple drug therapies assessed by means of network meta-analysis including only randomized controlled trials. Our results highlight that coadministrations have improved efficacy over monotherapy and the necessity of adapting current STH control strategies for the successful continuation of PC programs.

1. INTRODUCTION

Soil-transmitted helminths (STHs) – the roundworm (*Ascaris lumbricoides*), the whipworm (*Trichuris trichiura*), and the hookworms (*Ancylostoma duodenale* and *Necator americanus*) – affect more than 1.5 billion people worldwide with highest infection rates observed in marginalized population groups in the tropics and subtropics (Lustigman et al., 2012; Pullan et al., 2014). STH infections are responsible for the highest burden among the cluster of neglected tropical diseases (NTDs) with an estimated global burden of 3.3 million disability-adjusted life years (DALYs) in 2016 (GBD 2016 DALYs and HALE Collaborators, 2017). Chronic STH infections reduce the body's ability to absorb nutrients and vitamins, which in turn exacerbates malnutrition. Additionally, heavy infections have been associated with anaemia, stunted growth, and impaired intellectual development (Lustigman et al., 2012).

The global strategy to control morbidity due to STH infections is preventive chemotherapy (PC) with one of the two benzimidazole drugs - albendazole (ALB) and mebendazole (MEB) - that are, depending on prevalence, annually or biannually administered to school-aged children, at risk-populations including women of reproductive age and preschool-aged children (WHO, 2012). Both drugs are donated by the pharmaceutical industry to the World Health Organization (WHO), which is responsible for the coordination and overview of country programs (Crompton and WHO, 2006; McCarty et al., 2014). While ALB and MEB are both highly efficacious against *A. lumbricoides*, ALB shows a better efficacy profile against hookworms while both drugs show low efficacy against *T. trichiura* using single, oral doses. Alarmingly, based on a recent network meta-analysis considering time-interactions, the efficacy of both drugs is decreasing over time, which might be associated with anthelmintic drug resistance (Moser et al., 2017a).

Therapies combining two or more drugs are widely advocated in different therapeutic areas such as psychiatry, tuberculosis, malaria, cancer, HIV/AIDS or lymphatic filariasis (Cihlar and Fordyce, 2016; Fischer et al., 2017). The underlying rationale for multifactorial pharmacological treatment varies by disease and includes the protection against the development of drug resistance, hence a prolongation of the life-span of effective and available drugs, and an increase and broadening of the efficacy over the administration of drugs in monotherapy (Klastersky and Georgala, 2014).

2. SOIL-TRANSMITTED HELMINTHIASIS

2.1 Preventive Chemotherapy

In 2001, the World Health Assembly (World Health Assembly, 2001) endorsed PC as control strategy for all STH endemic countries with the aim of reaching 75% of school-aged children by 2010. After failing the initial goals, WHO revised their strategic plan for the control of STH infections and schistosomiasis in 2012. The updated goals aimed to target 50% of pre-and school-aged children in need for regular treatment by 2015 and up to 75% by 2020 with PC in order to eliminate STH as a public health problem (WHO, 2012). Inspired by the WHO roadmap, a collaborative control program (London Declaration on Neglected Tropical Diseases) was launched in the same year. Governments, global health organisations and industry endorsed their commitment to combat NTDs (Uniting to Combat NTDs, 2012), i.e., by increasing drug donations for the treatment of STH infections. Indeed, the latest numbers from 2016 indicate that 633 million pre-and school-aged children received treatment, which corresponds to a global PC coverage of 51% for preschool-aged children and 69% for school-aged children (WHO, 2017a). Despite the recent upscale of PC, its success is threatened by two major restrictions, first, the low efficacy of the current anthelmintics in particular against *T. trichiura* and second, the potential of anthelmintic drug resistance.

2.2 Anthelmintic Drug Resistance

The review by Moser and colleagues demonstrated a considerable global temporal decrease of the benzimidazole efficacy. For example, the egg reduction rate (ERR) estimates for the year 1995 and 2015 indicated a drop from 73% to 43% for ALB and from 91% to 55% for MEB against *T. trichiura*, respectively (Moser et al., 2017a). The decrease in efficacy could be explained by study related confounders or indeed by anthelmintic drug resistance.

With the up-scaling of PC over the past years, the selective drug pressure on STH populations increased, which potentially could trigger anthelmintic drug resistance (Vercruysse et al., 2011). While resistance in human STH has not been confirmed, in veterinary medicine frequent and rapid resistance selection against the major anthelmintic drugs including multi-drug resistance in strongyle parasites is a widely known problem (Abongwa et al., 2017; Kaplan, 2004; Kaplan and Vidyashankar, 2012). High annual treatment frequency, deworming over a longer time period, underdosing and small refugia are associated with resistance development and drug failure in animals (Geerts and Gryseels, 2000; Hoekstra et al., 1997). The smaller scale treatment programs might have prevented resistance until now in human medicine. While in veterinary practices 100% of animals are commonly dewormed, PC programs for humans target mainly pre-and school-aged children and at risk populations for the treatment of STH infections (WHO, 2017a). Hereby, the pool of wild type worms remains large and outcompetes resistant types, which might have prevented drug resistance in humans (Geerts and Gryseels, 2000).

A number of clinical trials reported reduced benzimidazole efficacy against hookworm and *T. trichiura* (Vercruysse et al., 2011) and more recently, reduced ALB efficacy against *A. lumbricoides* in Rwandan schoolchildren (Krücken et al., 2017). However, both studies had methodological weaknesses: one of them used a very short follow-up (7—10 days) for measuring the effect of *A. lumbricoides* treatment and both failed to associate reduced efficacy with the presence of a putative resistance marker following analysis of β-tubulin isotype genes. For *Haemonchus contortus*, a nematode of huge veterinary importance, benzimidazole resistance was associated with single nucleotide polymorphism (SNP) in the β-tubulin gene at three positions (codon 200, 167 and 198) (Ghisi et al., 2007). Studies investigating the presence of these SNPs in human STH reported a SNP in each species, i.e., in codon 167 for *A. lumbricoides* and 200 for hookworm and *T. trichiura* (Diawara et al., 2009, 2013b). After treatment with ALB an increased frequency of SNPs in codon 200 of *T. trichiura* was observed, however, an association with reduced efficacy could not be confirmed (Diawara et al., 2013a). All studies on the benzimidazole's mode of action and resistance development derive from *H. contortus*, which might not be directly translatable to human STH. Thus, further studies are required to elucidate fully the mode of action of benzimidazoles in human STHs, to identify the SNPs associated with drug resistance and to establish their potential as resistance markers for anthelmintic efficacy studies.

3. ALTERNATIVE DRUGS AND TREATMENT STRATEGIES

Considering the reduced efficacy of the current drugs and the lack of an effective drug particularly against *T. trichiura* (Moser et al., 2017a), research on new drug candidates is of pressing need. The main strategy in human anthelmintic drug discovery and development comprised the repurposing of old drugs from veterinary medicine, the reconsideration of previously marketed drugs and to a much smaller extent the discovery of new drugs (Olliaro et al., 2011).

Considering that only a handful drugs are currently on the market or in the anthelmintic drug pipeline, the coadministration of two drugs might be the way forward. Coadministration offers the advantage of an increased efficacy compared to monotherapy and by combining drugs, which act on different targets (Table 1), resistance might be delayed, which in the long term likely will outweigh the increased costs of such a treatment. In anthelmintic drug research, different combinations have been evaluated in several clinical trials mainly in the previous century, however, many trials were of exploratory nature, used questionable coadministrations and doses and none of the combinations was therefore introduced in PC programs (Keiser and Utzinger, 2010).

3.1 Tiers

In March 2016, the Bill and Melinda Gates Foundation (BMGF) hosted an expert meeting to review existing pharmacological and efficacy data on drug candidates which could be used in combination. To help the group prioritize candidate combinations, BMGF defined different levels of investment risk. In more detail, the prioritization includes four tiers and takes into account (i) the current efficacy and safety data of existing anthelmintic drugs; (ii) the financial and time investment required to generate the necessary evidence to change WHO treatment guidelines; and (iii) the potential for transformational change. Since ALB and ivermectin (IVR) are already on the WHO's Model List of Essential Medicines (MLEM, WHO 2017b), the first tier prioritizes the combination of these two drugs (Box 1). The

Table 1 Mode of Action of Anthelmintic Drugs

Class	Drug	Mode of Action	Effect	References
Benzimidazoles	ALB, MEB	Inhibition of microtubule polymerization by selective binding to the β-tubulin	Destruction of cell structure, death of parasite	(Lacey, 1990)
Avermectins	IVR	Glutamate-gated and gaba-amino-butyric-acid (GABA)-gated chloride and potassium channel agonist	Spastic paralysis of worm, expulsion of parasite and death	(Geary, 2005)
Milbemycin	MOX	Glutamate-gated and gaba-amino-butyric-acid (GABA)-gated chloride and potassium channel agonist	Spastic paralysis of worm, expulsion of parasite and death	(Cobb and Boeckh, 2009)
Imidazothiazoles	LEV	L-subtype nAChR agonist	Spastic paralysis of worm, expulsion of parasite	(Qian et al., 2006)
Tetrahydropyrimidines	PYP	L-subtype nAChR agonist	Spastic paralysis of worm, expulsion of parasite	(Robertson et al., 1994)
	OXP	N-subtype nAChR agonist	Spastic paralysis of worm, expulsion of parasite	(Martin et al., 2004)
Symmetrical diamidines	TRB	B-subtype nAChR agonist	Spastic paralysis of worm, expulsion of parasite	(Robertson et al., 2015)

Box 1 Drug combination tiers for the treatment of STH infections

Tier 1	Albendazole + ivermectin
Tier 2	Albendazole + oxantel pamoate (or oxantel pamoate-pyrantel pamoate)
	Mebendazole + oxantel pamoate (or oxantel pamoate-pyrantel pamoate)
	Mebendazole + ivermectin
Tier 3	Albendazole + moxidectin
	Albendazole + tribendimidine
	Moxidectin + tribendimidine
	Tribendimidine + oxantel pamoate (or oxantel pamoate-pyrantel pamoate)
	Tribendimidine + ivermectin
Tier 4	Oxfendazole + drug X
	Emodepside + drug X
	Cry5B + drug X

second tier includes combinations of the benzimidazoles with pyrantel pamoate (PYP), which is already on the WHO's MLEM and additionally, oxantel pamaote (OXP). The third tier entails the novel anthelmintics moxidectin (MOX) and tribendimidine (TRB) in combination with approved drugs and OXP. The fourth tier includes drugs under development with a potential effect against STHs as i.e., oxfendazole, emodepside and Cry5B, however, pre-and clinical trials have to confirm their potential.

4. EFFICACY OF ANTHELMINTIC TREATMENTS

In this publication, we present findings from a network meta-analysis including available evidence from clinical trials with single drugs, coadministrations and triple drug therapies (TDT), in order to investigate their potential for future PC programs. The database of the already published review and network meta-analysis on ALB, MEB, LEV and PYP (Moser et al., 2017) was extended by including the additional triple MeSH search terms "tribendimidine," "oxantel pamoate," "ivermectin," "moxidectin," "coadministration," "combination" and "triple drug therapy." We also updated our recent findings (Moser et al. 2017a) on the efficacy of the four currently recommended single drugs (ALB, MEB, levamisole (LEV) and PYP) which allows to strengthen the network analysis, since for some coadministrations and TDTs only little data are available. All randomized clinical trials including a single dose regimen or a combination of ALB (400 mg), MEB (500 mg), levamisole (LEV, 80 mg or 2.5 mg/kg), PYP (10–20 mg/kg), TRB (200 mg for children and 400 mg for adults), IVR (according to height 150–400 μg/kg), OXP (10–25 mg/kg), MOX (8 mg) and published before the 1st May 2018 were considered eligible. The analysis of the cure rates (CRs) were based on the method described in the previous network meta-analysis (Moser et al., 2017a,b) and details for the estimates and analyses of the ERRs are listed in the Supplementary Appendix Text S1.

4.1 Efficacy of the Currently Recommended Drugs

Anthelmintic drug efficacy is measured qualitatively by CRs or quantitatively by ERRs. The efficacy indicator CR only distinguishes the presence or absence of helminth eggs in stool. However, in the light of the current strategy of WHO to reduce the burden caused by STHs, the ERR - the reduction in egg output after treatment – is the key indicator for anthelmintic drug efficacy (WHO, 2013). The four drugs ALB, MEB, LEV, and PYP are on the WHO's MLEM against intestinal helminths (WHO, 2015). The first summary estimates of their efficacy was published a decade ago (Keiser and Utzinger, 2008). Moreover, in the past 2 years, four independent reviews presented summary estimates of at least one of the recommended drugs (Levecke and Vercruysse, 2016; Moser et al., 2017a; Mrus et al., 2017; WHO, 2017b). The efficacy of the four recommended drugs from the five past reviews and from the present publication are presented in Table 2. The statistics applied by these reviews ranged from simple summary estimates (mean or median (Levecke and Vercruysse, 2016; Mrus et al., 2017; WHO, 2017b), over a meta-analysis for CRs (Keiser and Utzinger, 2008), up to a network meta-analysis for CRs and ERRs ((Moser et al., 2017a) and present publication).

As to *A. lumbricodes*, the summary estimates show similar efficacy, with CRs of at least 88% and ERRs exceeding 94% for all drugs (Table 1). ALB reveals highest efficacy against hookworm, with a CR between 72% and 80% and an ERR between 90% and 92%. The estimates for MEB

Table 2 The Efficacy (Cure Rate and Egg Reduction Rate) for ALB, MEB and PYP According to all Reviews and Meta-analyse

Review	ALB		MEB		PYP		LEV	
	CR	ERR	CR	ERR	CR	ERR	CR	ERR
Ascaris lumbricoides								
Keiser and Utzinger (2008) [a]	88.0	—	95.0	—	88.0	—	—	—
Levecke and Vercruysse 2016 [b]	—	—	—	—	96.0	98.5	—	—
Mrus et al. (2017) [a] ([b])	—	—	84.6 (92.6)	98.1 (97.9)	—	—	—	—
WHO (2017b) [b] ([c])	—	98.7 (99.9)	—	98.3 (97.6)	—	—	—	—
Moser et al. 2017b	95.7	98.5	96.2	98.0	92.6	94.3	97.3	96.4
Present publication (Table 3)	96.5	99.7	96.8	99.5	93.0	97.0	97.5	91.7
Hookworm								
Keiser and Utzinger (2008) [a]	72.0	—	15.0	—	31.0	—	—	—
Levecke and Vercruysse (2016) [b]	—	—	—	—	54.5	67.0	—	—
Mrus et al. (2017) [a] ([b])	—	—	25.5 (20.5)	72.0 (61.2)	—	—	—	—
WHO (2017b) [b] ([c])	—	89.8 (92.4)	—	68.2 (76.5)	—	—	—	—
Moser et al. 2017b	79.5	89.6	32.5	61.0	49.8	71.9	10.3	61.8
Present publication (Table 4)	78.5	92.1	41.6	65.1	52.6	80.4	14.2	65.3
Trichuris trichiura								
Keiser and Utzinger (2008) [a]	28.0	—	36.0	—	—	—	—	—
Levecke and Vercruysse (2016) [b]	—	—	—	—	19.0	48.0	—	—
Mrus et al. (2017) [a] ([b])	—	—	27.6 (33.4)	72.5 (86.8)	—	—	—	—
WHO (2017b) [b] ([c])	—	60.7 (64.4)	—	69.0 (69.3)	—	—	—	—
Moser et al. 2017b	30.7	49.9	42.1	66.0	20.2	47.5	29.5	28.3
Present publication (Table 5)	32.1	64.3	44.4	80.7	23.4	41.8	28.5	62.3

[a]only placebo-controlled trials.
[b]all studies.
[c]WHO methodology.

(CR: 15%–42% and ERR: 61%–77%) and PYP (CR: 31%–55% and ERR: 67%–80%) show higher variability among the estimates depending on the review methodology. Overall, none of the drugs demonstrated high efficacy against *T. trichiura*, with moderate efficacy for MEB (CR: 28%–44% and ERR: 66%–87%) and only marginal efficacies for either ALB (CR: 28% and 32%, ERR: 50%–64%), PYP (CR: 19%–23% and ERR 41%–48%) or LEV (CR: 29%–30%, ERR: 28%–62%).

4.2 Alternative Single Drugs, Coadministrations and Triple Drug Therapies

4.2.1 Tribendimidine

In the mid-1980s, TRB was developed as a novel broad spectrum anthelmintic drug in China. A series of preclinical and clinical trials led to the approval of TRB as a human anthelmintic by the Chinese Food and Drug Administration in 2004 (Xiao et al., 2005). TRB is a B-subtype selective nAChR agonist (Robertson et al., 2015) and active against more than 20 different parasite species including nematodes (e.g., *Enterobius vermicularis, A. lumbricoides* and both hookworms (Wu et al., 2006; Zhou et al., 2008)), cestodes (e.g., *Taenia* spp. (Steinmann et al., 2008)) and trematodes (*Clonorchis sinensis* (Qian et al., 2013)) and *Opistorchis viverrini* (Sayasone et al., 2016; Soukhathammavong et al., 2011). Throughout all clinical studies, TRB has demonstrated an excellent safety profile (Xiao et al., 2013). The drug was well tolerated and patients showed no abnormalities in ECG examinations and routine blood, urine, hepatic and renal functions (Xiao et al., 2005). Moreover, all studies reported only low numbers of adverse events for the recommended single dose treatment (200 mg for children and 400 mg for adults) (Xiao et al., 2013).

We included six studies (Cao et al., 2000; Fang et al., 2002; Moser et al., 2017b; Steinmann et al., 2008; Xiao et al., 2005; Xu et al., 2014) to estimate the efficacy of TRB, which is comparable to ALB. In more detail, the efficacy of TRB against hookworm (CR: 83.4%, 68.7–92.0, ERR: 91.6, 82.6–96.8, Table 4) was significantly higher (all $P < 0.05$, Appendix Table S2) compared to all monotherapies except ALB. Furthermore, TRB reached high efficacy against *A. lumbricoides* (CR: 96.5%, 79.2–99.7, ERR: 88.1%, 27.8–98.8, Table 3) but only low efficacy against *T. trichiura* (CR: 5.9%, 1.2–24.8, ERR: 41.3%, 3.2–72.0, Table 5).

4.2.2 Ivermectin

IVR was approved first for animal and in 1988 for human use (Campbell, 1991). It derived from avermectin (lactones) and is a semisynthetic

Table 3 Estimates of Cure Rates and Egg Reduction Rates Against *A. lumbricoides* for Single Drugs, Coadministrations and Triple Drug Therapies

	Cure Rate		Egg Reduction Rate	
Drugs	**%**	**95%–CI**	**%**	**95%–CI**
PLC	12.5	6.8–21.9	39.0	−4.3–73.1
ALB	96.5	94.4–97.9	99.7	99.2–99.9
MEB	96.8	93.3–98.5	99.5	98.5–99.9
LEV	97.5	82.8–99.7	91.7	−8.3–99.8
PYP	93.0	86.7–96.4	97.0	87.7–99.3
TRB	96.5	79.2–99.7	88.1	27.8–98.8
IVR	97.3	85.7–99.5	>99.9	99.9–99.9
OXP	21.8	6.9–51.2	35.8	−36.0–88.5
MOX	90.5	59.5–98.4	97.9	72.3–99.9
ALB-IVR	96.7	87.6–99.2	>99.9	99.8–99.9
ALB-OXP	95.2	85.7–98.4	98.3	90.9–99.7
PYP-OXP	95.3	87.2–98.4	99.9	99.3–>99.9
TRB-IVR	98.9	79.2–100.0	99.3	77.2–>99.9
TRB-OXP	94.2	56.1–99.5	99.3	77.2–>99.9
ALB-MOX	92.9	53.3–99.3	99.1	69.6–>99.9
TRB-MOX	93.8	55.6–99.5	99.1	69.6–>99.9
ALB-PYP-OXP	90.4	36.3–99.4	98.3	57.2–>99.9
MEB-PYP-OXP	100.0[a]		100.0[a]	

[a]observed CR and ERR because of small sample size no analysis was possible. PLC, Placebo.

macrocyclic lactone. Similar to MOX, IVR is a selective agonist of the glutamate-gated ion channels and has an activity on the Gaba-amino-butyric-acid (GABA)-gated chloride and potassium channels, which is causing paralysis of the parasites (Campbell, 1991; Geary, 2005). Currently, IVR is used worldwide against onchocerciasis, strongyloidiasis, scabies, pediculosis, gnathostomiasis and myiasis (Santos et al., 2009) and is since recently on the WHO's MLEM in combination with albendazole against STH infections (WHO, 2017c). IVR is considered as a very safe drug, however, serious adverse events have been reported from *Loa loa* coinfected people (Gardon et al., 1997) and thus, IVR is not recommended in *L. loa* endemic countries (WHO, 2017d).

A total of eight clinical trials (Barda et al., 2017; Belizario et al., 2003; Marti et al., 1996; Naquira et al., 1989; Ndyomugyenyi et al., 2008; Opoku et al., 2018; Wen et al., 2008; Wimmersberger et al., 2018) were eligible for estimating the efficacy of IVR. Overall high efficacy was reached against *A. lumbricoides* (CR: 97.3%, 85.7–99.5, ERR: >99.9, 99.9–>99.9,

Table 4 Estimates of Cure Rates and Egg Reduction Rates Against Hookworm (Both Species) for Single Drugs, Coadministrations and Triple Drug Therapies

	Cure Rate		Egg Reduction Rate	
Drugs	**%**	**95%–CI**	**%**	**95%–CI**
PLC	16.2	10.7–23.6	22.1	−0.4–41.2
ALB	78.5	71.5–84.2	92.1	88.5–94.8
MEB	41.6	30.7–53.3	65.1	50.8–75.8
LEV	14.2	4.4–37.2	65.3	4.7–90.7
PYP	52.6	39.4–65.4	80.4	68.1–88.6
TRB	83.4	68.7–92.0	91.6	82.6–96.8
IVR	24.3	11.5–44.3	76.3	54.1–89.0
OXP	23.8	9.5–48.2	39.5	−6.0–69.1
MOX	37.6	17.1–63.9	76.0	48.9–90.3
ALB-IVR	83.7	68.1–92.6	94.7	88.0–98.6
ALB-OXP	72.0	53.5–85.1	91.3	82.7–96.4
PYP-OXP	82.8	66.7–92.1	87.7	74.9–94.9
TRB-IVR	93.5	73.3–98.7	99.5	91.0–>99.9
TRB-OXP	73.9	35.7–93.5	88.8	59.3–98.8
ALB-MOX	76.7	38.1–94.6	95.7	78.8–>99.9
TRB-MOX	88.5	55.7–97.9	96.5	81.5–>99.9
ALB-PYP-OXP	92.8	70.7–98.6	96.7	82.3–>99.9
MEB-PYP-OXP	84.7	50.6–96.8	93.4	72.2–>99.9

PLC, Placebo.

Table 3), whereas low CRs and moderate ERRs were estimated against hookworm (CR: 24.3%, 11.5–44.3, ERR: 76.3%, 54.1–89.0, Table 4) and *T. trichiura* (CR: 32.1%, 16.3–53.5; ERR: 78.9%, 60.3–90.0, Table 5).

4.2.3 Oxantel Pamoate

OXP combined with PYP was developed and marketed under the brand name Quantrel® (Pfizer) for treating human helminth infections in several countries, but is no longer produced (Kopp and Keiser, 2017). OXP belongs to the tetrahydropyrimidines and is selectively gating the N-subtype of the nicotinic acetylcholine receptors (nAChRs) in nerves (Table 2), which is causing spastic paralysis of the nematodes muscles and results in the worm's expulsion (Martin et al., 2004). OXP is water soluble and poorly absorbed from the human gastrointestinal tract, which might explain that only few and mild adverse events were described in several studies (Garcia, 1976; Lee et al., 1976; Moser et al., 2016, 2017b; Peldán and Pitkänen, 1982; Sinniah et al., 1980; Speich et al., 2014, 2015).

Table 5 Estimates of Cure Rates and Egg Reduction Rates Against *T. trichiura* for Single Drugs, Coadministrations and Triple Drug Therapies

	Cure Rate		Egg Reduction Rate	
Drugs	**%**	**95%–CI**	**%**	**95%–CI**
PLC	10.6	6.3–17.2	31.2	12.5–49.0
ALB	32.1	23.2–42.5	64.3	53.2–73.5
MEB	44.4	30.8–59.0	80.7	70.6–87.8
LEV	28.5	7.7–65.7	62.3	4.4–91.5
PYP	23.4	11.7–41.1	41.8	9.0–68.8
TRB	5.9	1.2–24.8	41.3	3.2–72.0
IVR	32.1	16.3–53.5	78.9	60.3–90.0
OXP	75.7	59.2–87.0	85.0	73.7–92.1
MOX	30.1	8.2–67.6	81.0	50.7–94.4
ALB-IVR	60.0	43.3–74.6	95.5	88.7–98.8
ALB-OXP	88.7	76.2–95.1	96.7	91.9–99.2
PYP-OXP	63.3	46.2–77.6	85.1	74.7–91.8
TRB-IVR	35.4	8.2–77.1	91.2	63.0–99.2
TRB-OXP	68.4	26.0–93.0	95.6	77.6–>99.9
ALB-MOX	65.6	24.7–91.7	92.2	65.7–99.5
TRB-MOX	34.9	8.2–76.3	79.9	34.3–96.4
ALB-PYP-OXP	84.2	39.7–97.7	92.7	68.8–99.5
MEB-PYP-OXP	75.2	26.9–96.2	90.7	62.9–99.0

PLC, Placebo.

OXP is of particular interest due to its unprecedented high activity against *T. trichiura*. Based on data from six randomized controlled studies (Choi et al., 1979; Garcia, 1976; Lee et al., 1976; Lim, 1978; Moser et al., 2016; Speich et al., 2014), the CR of OXP monotherapy (CR 75.7%, 59.2–87.0) was significantly higher compared to all remaining monotherapies and the ERR (85.0%, 73.7–92.1) in comparison with ALB, PYP and TRB (Table 5, Appendix Table S3). Only marginal effects against *A. lumbricoides* and hookworm were estimated (Tables 3 and 4). Moreover, a recent dose-ranging trial investigated the efficacy of different dose regimens (5–30 mg/kg) and identified peak efficacy for 25 mg/kg (CR: 60%, ERR: 98%) and recommended a weight-independent dose of 500 mg for school-aged children (Moser et al., 2016).

4.2.4 Moxidectin

Moxidectin was developed for the veterinary market and is in use since 1985 (Bowman et al., 1988). Today, MOX is marketed by several pharmaceutical companies as treatment against parasites in cattle, sheep, horses and dogs

(Lyons and Tolliver, 2012; Nolan and Lok, 2012). It belongs to the milbemycins and is a broad-spectrum endectocide. The drug is currently developed for the treatment of human onchocerciasis by the Australian not-for-profit organisation 'Medicine Development for Global Health' (Kuesel, 2016). Moxidectin belongs to the second-generation of macrocyclic lactones, which bind to gamma-aminobutyric (GABA) and glutamate-gated chloride and potassium channels. Despite its similar mode of action compared to IVR, both act on different subunits and MOX is still effective against IVR resistant strains (Bygarski et al., 2014; Craig et al., 1992). The binding on the channels increases the permeability and therewith the influx of ions, which results in the paralysis and death of the parasites (Cobb and Boeckh, 2009). Studies in rats and dogs indicated a very low absorption of MOX, low plasma clearance and high volume of distribution, resulting in a long half-life (Cotreau et al., 2003; Korth-Bradley et al., 2012).

The efficacy estimates for MOX derive from three studies (Barda et al., 2017, 2018; Opoku et al., 2018) and indicate its high efficacy against *A. lumbricoides* (CR: 90.5%, 59.5−98.4, ERR: 97.9%, 72.3−99.9, Table 3), but moderate efficacy against hookworm (CR: 37.6%, 17.1−63.9, ERR: 76.0%, 48.9−90.3, Table 4) and *T. trichiura* (CR: 30.1%, 8.2−67.6, ERR: 81.0%, 50.7−94.4, Table 5).

4.2.5 Efficacy of Coadministrations and Triple Drug Therapies

We included data from 17 clinical trials of drug combinations in our network meta-analysis to estimate the efficacy of the coadministrations ALB-IVR (Beach et al., 1999; Belizario et al., 2003; Knopp et al., 2010; Ndyomugyenyi et al., 2008; Speich et al., 2015), ALB-OXP (Barda et al., 2018; Moser et al., 2017b, 2018, Speich et al., 2014, 2015), PYP-OXP (Albonico et al., 2002; Cabrera et al., 1980; Cabrera and Cruz, 1980; Choi et al., 1979; Lee and Lim, 1978; Lim, 1978; Moser et al., 2018; Rim et al., 1981; Sinniah and Sinniah, 1981), TRB-IVR (Moser et al., 2017b), TRB-OXP (Moser et al., 2017a), ALB-MOX (Barda et al., 2018), TRB-MOX (Barda et al., 2018), as well as the two triple drug therapies (TDTs) ALB-PYP-OXP (Moser et al., 2018) and MEB-PYP-OXP (Moser et al., 2018).

For treating *A. lumbricoides* infections, the estimated CRs and ERRs of all single drugs, coadministrations and TDT reached 90% or above (Table 3). The highest cure rate was reached in the clinical trial with MEB-PYP-OXP where all participants were cured, followed by TRB-IVR (CR: 98.9%, 79.2−100.0, ERR 99.3%, 77.2−>99.9). IVR was significantly more efficacious in terms of ERRs compared to all other single drug, coadministrations and the TDT including ALB (all $P < 0.05$, Appendix Table S1). Furthermore, the ERR of ALB-IVR (>99.9%), was significantly higher ($p < 0.05$) than the ERR's of all single drugs with exception of IVR.

Among the treatments against hookworm, the efficacy reached by the coadministrations and TDTs were generally higher compared to the single drugs – except for ALB and TRB – with overall peak efficacy for TRB-IVR (CR: 93.5%, 73.3–98.7, ERR: 99.5%, 91.0–>99.9, Table 4). Similar high efficacy was estimated for ALB-PYP-OXP (CR: 92.8, 70.7–98.6; ERR: 96.7%, 82.3–>99.9), TRB-MOX (CR: 88.5%, 55.7–97.9; ERR: 96.5%, 81.5–>99.9) and ALB-IVR (CR 83.7%, 68.1–92.6, ERR 94.7%, 88.0–98.6).

For treating *T. trichiura* infections, all coadministrations and TDTs but only OXP as monotherapy reached moderate to high efficacy (Table 5). The greatest efficacy was calculated for the coadministration ALB-OXP (CR 88.7%, 76.2–95.1, ERR 96.7%, 91.9–99.2), the respective estimates being significantly higher than those of all single drugs except OXP and of the coadministrations PYP-OXP and TRB-MOX (Appendix Table S3). Furthermore, ALB-PYP-OXP (CR 84.2%, 39.7–97.7, ERR 92.7%, 68.8–99.5), TRB-OXP (CR 68.4%, 26.0–93.0, ERR 95.6%, 77.6–>99.9) and ALB-IVR (CR 60.0%, 43.3–74.6; ERR 95.5%, 88.7–98.8) ranked high among the treatment options analysed revealing considerable efficacy.

4.3 Ranking of all Treatment Options

Common network meta-analysis allows to rank the efficacy of the included treatments (Chaimani et al., 2013). Since two measurements are commonly used (CR and ERR) for anthelmintic drug efficacy, a different ranking approach was applied. The rank of each treatment was obtained by first taking the average value of the CR and double the value of ERR, since ERRs are the current key estimate for anthelmintic drug efficacy (WHO, 2013), The ranking, whose results are presented in Fig. 1, was performed for each helminth species separately and for all species combined.

Considering the combined ranking, the TDT ALB-PYP-OXP showed highest efficacy to treat all three STH species simultaneously. The coadministration ALB-OXP was ranked second and the TDT containing MEB third. The review results show that all coadministrations and both TDTs were ranked above the single drugs and therewith outcompete all monotherapies as treatment against STH. Even when considering only individual species,

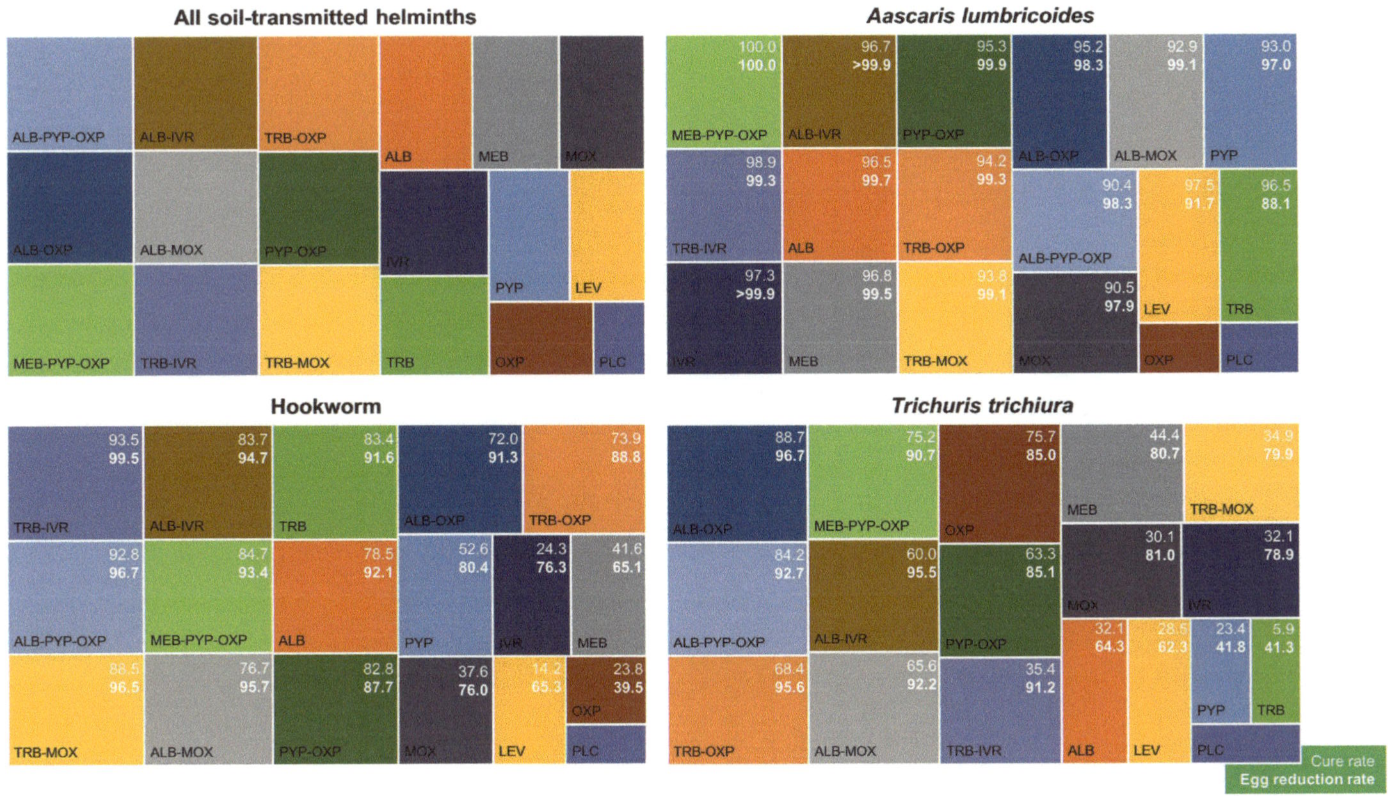

Figure 1 All single drugs, coadministrations and triple drugs therapies were ranked once for *A. lumbricoides*, hookworm, *T. trichiura* and for all three STH species together. The treatment on top left ranks first with highest efficacy with descending efficacy from top to bottom and left to right and overall lowest efficacy for the treatment on the bottom right.

highest efficacies were determined for the coadministrations and TDTs. In more detail, the TDT with MEB and TRB-IVR were the two most effective treatments against *A. lumbricoides*, followed by IVR and ALB-IVR. With regard to hookworm, TRB-IVR ranked above ALB-PYP-OXP and TRB-MOX. A similar ranking was reached for *T. trichiura*, with peak efficacy for ALB-OXP, and ALB-PYP-OXP and TRB-OXP coming next.

5. CONCLUSIONS AND RESEARCH GAPS

About 1.5 billion people (Pullan et al., 2014) are infected with at least one of the STH species, yet research on and development of new drugs to replace the widely used ALB and MEB is progressing only very slowly. Solely one single new drug (TRB) received the approval for human use in the past 3 decades (Xiao et al., 2005) and registration at a stringent regulatory authority is still pending. Moreover, the two drugs PYP and LEV listed since almost half a century on WHO's MLEM (WHO, 2015), are presently not considered for PC programs since there are no donation plans for these drugs (Brooker, 2018). Since the goal of WHO is to reduce the STH burden to a level which is no longer considered a public health problem by 2020 and over 1 billion tablets of either ALB or MEB are distributed every year in PC programs (WHO, 2012, 2017a), an enormous drug pressure is exerted on the helminths. Although resistance in human medicine is not yet confirmed, its manifestation might only be a matter of time. Moreover, the low efficacy of ALB against *T. trichiura* and of MEB against hookworm and *T. trichiura* is a reality and lowers the effectiveness of PC programs.

This review demonstrated: (i) alternative treatments are available with high efficacy against hookworm (TRB) and *T. trichiura* (OXP); (ii) coadministrations have increased efficacy against all three STHs when compared to monotherapy and (iii) the combination of three drugs performed much better than any other treatment option. Since the coadministration of ALB-IVR is on WHO's MLEM, donated by the respective pharmaceutical companies, distributed already widely in the framework of lymphatic filariasis control programs (WHO, 2017d) and has a broad efficacy spectrum, ALB-IVR should be added to STH PC programs. However, based on our final ranking, the coadministration ALB-OXP showed higher efficacy than ALB-IVR, which was mainly attributed to the lower efficacy against *T. trichiura* of the latter. Thus, ALB-OXP is the most promising coadministration available.

This coadministration is listed in the BMGF tier 2, together with the two TDTs (ALB-PYP-OXP and MEB-PYP-OXP). Considering lymphatic filariasis elimination programs, a TDT (IVR-diethylcarbamazine-ALB) has recently been recommended for large scale use (WHO, 2017d). Hence, TDTs might gain importance in the field of STHs, once the strategy is moving from STH control to elimination (Moser et al., 2018). However, there is a need to register OXP at a stringent regulatory authority and make the drug commercially available. Moreover, donation plans for both OXP and PYP need to be established. Tier 3, includes the new drugs TRB and MOX, whereof TRB might be important to prevent resistance against ALB or MEB (Moser et al., 2017b). Dose-optimization studies of MOX against *T. trichiura* have been launched and there is a need to optimize the drug for STH infections. TRB in combination with IVR or MOX could be of particular interest for hookworm endemic areas, since both coadministrations showed indications of synergism in three clinical trials (Barda et al., 2018; Moser et al., 2017b; Wu and Qian, 2003).

In summary, when used in appropriate coadministrations, even with the few available anthelmintic drugs, STH infections can be safely and effectively treated. Combined efforts between researchers, public health authorities, industry and local governments are now needed to tackle the huge undertaking of developing and deploying these treatments so that STH can be effectively controlled or even eliminated in the not so far future.

ACKNOWLEDGEMENTS

We are grateful to Dr Bruno Levecke for providing his database including all studies with pyrantel pamoate and oxantel pamoate.

APPENDIX A SUPPLEMENTARY DATA

Supplementary data related to this article can be found at https://doi.org/10.1016/bs.apar.2018.08.002

REFERENCES

Abongwa, M., Martin, J., Robertson, A., 2017. A brief review on the mode of action of antinematodal drugs: Acta Vet. (Beogr.) 67, 137–152. https://doi.org/10.1515/acve-2017-0013.

Albonico, M., Bickle, Q., Haji, H.J., Ramsan, M., Khatib, K.J., Montresor, A., Savioli, L., Taylor, M., 2002. Evaluation of the efficacy of pyrantel-oxantel for the treatment of soil-transmitted nematode infections. Trans. R. Soc. Trop. Med. Hyg. 96, 685–690. https://doi.org/10.1016/S0035-9203(02)90352-4.

Barda, B., Ame, S.M., Ali, S.M., Albonico, M., Puchkov, M., Huwyler, J., Hattendorf, J., Keiser, J., 2018. Efficacy and tolerability of moxidectin alone and in co-administration with albendazole and tribendimidine, and oxantel pamoate-albendazole against *Trichuris trichiura* infections: a randomized, non-inferiority, single blind trial. Lancet Infect. Dis. 18, 864–873. https://doi.org/10.1016/S1473-3099(18)30233-0.

Barda, B., Sayasone, S., Phongluxa, K., Xayavong, S., Keoduangsy, K., Odermatt, P., Puchkov, M., Huwyler, J., Hattendorf, J., Keiser, J., 2017. Efficacy of moxidectin versus ivermectin against *Strongyloides stercoralis* infections: a randomized controlled non-inferiority trial. Clin. Infect. Dis. 65, 276–281. https://doi.org/10.1093/cid/cix278.

Beach, M.J., Streit, T.G., Addiss, D.G., Prospere, R., Roberts, J.M., Lammie, P.J., 1999. Assessment of combined ivermectin and albendazole for treatment of intestinal helminth and *Wuchereria bancrofti* infections in Haitian schoolchildren. Am. J. Trop. Med. Hyg. 60, 479–486.

Belizario, V.Y., Amarillo, M.E., de Leon, W.U., de los Reyes, A.E., Bugayong, M.G., Macatangay, B.J., 2003. A comparison of the efficacy of single doses of albendazole, ivermectin, and diethylcarbamazine alone or in combinations against *Ascaris* and *Trichuris* spp. Bull. World Health Organ. 81, 35–42. https://doi.org/10.1590/S0042-96862003000100008.

Bowman, D.D., Parsons, J.C., Grieve, R.B., Hepler, D.I., 1988. Effects of milbemycin on adult *Toxocara canis* in dogs with experimentally induced infection. Am. J. Vet. Res. 49, 1986–1989.

Brooker, S.J., 2018. Soil-transmitted helminth treatment: multiple-drug regimens. Lancet Infect. Dis. 18, 698–699. https://doi.org/10.1016/S1473-3099(18)30268-8.

Bygarski, E.E., Prichard, R.K., Ardelli, B.F., 2014. Resistance to the macrocyclic lactone moxidectin is mediated in part by membrane transporter P-glycoproteins: implications for control of drug resistant parasitic nematodes. Int. J. Parasitol. Drugs Drug Resist. 4, 143–151. https://doi.org/10.1016/j.ijpddr.2014.06.002.

Cabrera, B.D., Cruz, A.C., 1980. Clinical trial of oxantel-pyrantel (Quantrel) against trichuriasis. Acta Med. Philipp. 16, 95–102.

Cabrera, B.D., Valdez, E.V., Go, T.G., 1980. Clinical trials of broad spectrum anthelmintics against soil-transmitted helminthiasis. Southeast Asian J. Trop. Med. Publ. Health 11, 502–506.

Campbell, W.C., 1991. Ivermectin as an antiparasitic agent for use in humans. Annu. Rev. Microbiol. 45, 445–474. https://doi.org/10.1146/annurev.mi.45.100191.002305.

Cao, H., Sun, F., Qian, Y., Chen, J., Zhao, L., Xu, Z., Wu, Z., 2000. Clinical observation of the efficacy of tribendimidine against hookworm infections. Chin. J. Parasit. Dis. Control 13, 184–186.

Chaimani, A., Higgins, J.P.T., Mavridis, D., Spyridonos, P., Salanti, G., 2013. Graphical tools for network meta-analysis in STATA. PLoS One 8. https://doi.org/10.1371/journal.-pone.0076654 e76654.

Choi, W.Y., Lee, O.R., Lee, W.K., Kim, W.K., Chung, C.S., Ough, B.O., 1979. A clinical trial of oxantel and pyrantel against intestinal nematodes infections. Kor. J. Parasitol. 17, 60. https://doi.org/10.3347/kjp.1979.17.1.60.

Cihlar, T., Fordyce, M., 2016. Current status and prospects of HIV treatment. Curr. Opin. Virol. 18, 50–56. https://doi.org/10.1016/j.coviro.2016.03.004.

Cobb, R., Boeckh, A., 2009. Moxidectin: a review of chemistry, pharmacokinetics and use in horses. Parasit. Vectors 2, S5. https://doi.org/10.1186/1756-3305-2-S2-S5.

Cotreau, M.M., Warren, S., Ryan, J.L., Fleckenstein, L., Vanapalli, S.R., Brown, K.R., Rock, D., Chen, C.Y., Schwertschlag, U.S., 2003. The antiparasitic moxidectin: safety, tolerability, and pharmacokinetics in humans. J. Clin. Pharmacol. 43, 1108–1115. https://doi.org/10.1177/0091270003257456.

Craig, T.M., Hatfield, T.A., Pankavich, J.A., Wang, G.T., 1992. Efficacy of moxidectin against an ivermectin-resistant strain of *Haemonchus contortus* in sheep. Vet. Parasitol. 41, 329–333.

Crompton, D.W.T., WHO, 2006. Preventive chemotherapy in human helminthiasis: Coordinated use of anthelminthic drugs in control interventions: a manual for health professionals and programme managers. Geneva World Health Organization.

Diawara, A., Drake, L.J., Suswillo, R.R., Kihara, J., Bundy, D.A.P., Scott, M.E., Halpenny, C., Stothard, J.R., Prichard, R.K., 2009. Assays to detect beta-tubulin codon 200 polymorphism in *Trichuris trichiura* and *Ascaris lumbricoides*. PLoS Negl. Trop. Dis. 3, e397. https://doi.org/10.1371/journal.pntd.0000397.

Diawara, A., Halpenny, C.M., Churcher, T.S., Mwandawiro, C., Kihara, J., Kaplan, R.M., Streit, T.G., Idaghdour, Y., Scott, M.E., Basáñez, M.G., Prichard, R.K., 2013a. Association between response to albendazole treatment and β-tubulin genotype frequencies in soil-transmitted helminths. PLoS Negl. Trop. Dis. 7, e2247. https://doi.org/10.1371/journal.pntd.0002247.

Diawara, A., Schwenkenbecher, J.M., Kaplan, R.M., Prichard, R.K., 2013b. Molecular and biological diagnostic tests for monitoring benzimidazole resistance in human soil-transmitted helminths. Am. J. Trop. Med. Hyg. 88, 1052–1061. https://doi.org/10.4269/ajtmh.12-0484.

Fang, Y.Y., Liang, W., Zhang, Q., Zhu, G., Liu, E., Liao, Z.Y., Huang, G., 2002. Study on the efficacy of tribendimidine for treatment of intestinal nematode infections. Chin. J. Schistosomiasis Control 02.

Fischer, P.U., King, C.L., Jacobson, J.A., Weil, G.J., 2017. Potential value of triple drug therapy with ivermectin, diethylcarbamazine, and albendazole (IDA) to accelerate elimination of lymphatic filariasis and onchocerciasis in Africa. PLoS Negl. Trop. Dis. 11. https://doi.org/10.1371/journal.pntd.0005163 e0005163.

Garcia, E.G., 1976. Treatment for trichuriasis with oxantel. Am. J. Trop. Med. Hyg. 25, 914–915.

Gardon, J., Gardon-Wendel, N., Demanga-Ngangue, null, Kamgno, J., Chippaux, J.P., Boussinesq, M., 1997. Serious reactions after mass treatment of onchocerciasis with ivermectin in an area endemic for *Loa loa* infection. Lancet Lond. Engl. 350, 18–22. https://doi.org/10.1016/S0140-6736(96)11094-1.

GBD 2016 DALYs and HALE Collaborators, 2017. Global, regional, and national disability-adjusted life-years (DALYs) for 333 diseases and injuries and healthy life expectancy (HALE) for 195 countries and territories, 1990-2016: a systematic analysis for the Global Burden of Disease Study 2016. Lancet Lond. Engl. 390, 1260–1344. https://doi.org/10.1016/S0140-6736(17)32130-X.

Geary, T.G., 2005. Ivermectin 20 years on: maturation of a wonder drug. Trends Parasitol. 21, 530–532. https://doi.org/10.1016/j.pt.2005.08.014.

Geerts, S., Gryseels, B., 2000. Drug resistance in human helminths: current situation and lessons from livestock. Clin. Microbiol. Rev. 13, 207–222. https://doi.org/10.1128/CMR.13.2.207-222.2000.

Ghisi, M., Kaminsky, R., Mäser, P., 2007. Phenotyping and genotyping of *Haemonchus contortus* isolates reveals a new putative candidate mutation for benzimidazole resistance in nematodes. Vet. Parasitol. 144, 313–320. https://doi.org/10.1016/j.vetpar.2006.10.003.

Hoekstra, R., Visser, A., Wiley, L.J., Weiss, A.S., Sangster, N.C., Roos, M.H., 1997. Characterization of an acetylcholine receptor gene of *Haemonchus contortus* in relation to levamisole resistance. Mol. Biochem. Parasitol. 84, 179–187.

Kaplan, R.M., 2004. Drug resistance in nematodes of veterinary importance: a status report. Trends Parasitol. 20, 477–481. https://doi.org/10.1016/j.pt.2004.08.001.

Kaplan, R.M., Vidyashankar, A.N., 2012. An inconvenient truth: global worming and anthelmintic resistance. Vet. Parasitol. 186, 70–78. https://doi.org/10.1016/j.vetpar.2011.11.048.

Keiser, J., Utzinger, J., 2010. The drugs we have and the drugs we need against major helminth infections. Adv. Parasitol. 73, 197–230. https://doi.org/10.1016/S0065-308X(10)73008-6.

Keiser, J., Utzinger, J., 2008. Efficacy of current drugs against soil-transmitted helminth infections: systematic review and meta-analysis. JAMA, J. Am. Med. Assoc. 299, 1937–1948. https://doi.org/10.1001/jama.299.16.1937.

Klastersky, J., Georgala, A., 2014. Strategies for the empirical management of infection in cancer patients with emphasis on the emergence of resistant gram-negative bacteria. Crit. Rev. Oncol. Hematol. 92, 268–278. https://doi.org/10.1016/j.critrevonc.2014.06.002.

Knopp, S., Mohammed, K.A., Speich, B., Hattendorf, J., Khamis, I.S., Khamis, A.N., Stothard, J.R., Rollinson, D., Marti, H., Utzinger, J., 2010. Albendazole and mebendazole administered alone or in combination with ivermectin against *Trichuris trichiura*: a randomized controlled trial. Clin. Infect. Dis. 51, 1420–1428. https://doi.org/10.1086/657310.

Kopp, S., Keiser, J., 2017. Pyrantel and oxantel pamoate. In: Kucers' The Use of Antibiotics: A Clinical Review of Antibacterial, Antifungal, Antiparasitic, and Antiviral Drugs, Seventh Edition. CRC Press, 3, pp. 3381–3384.

Korth-Bradley, J.M., Parks, V., Patat, A., Matschke, K., Mayer, P., Fleckenstein, L., 2012. Relative bioavailability of liquid and tablet formulations of the antiparasitic moxidectin. Clin. Pharmacol. Drug Dev. 1, 32–37. https://doi.org/10.1177/2160763X11432508.

Krücken, J., Fraundorfer, K., Mugisha, J.C., Ramünke, S., Sifft, K.C., Geus, D., Habarugira, F., Ndoli, J., Sendegeya, A., Mukampunga, C., Bayingana, C., Aebischer, T., Demeler, J., Gahutu, J.B., Mockenhaupt, F.P., von Samson-Himmelstjerna, G., 2017. Reduced efficacy of albendazole against *Ascaris lumbricoides* in Rwandan schoolchildren. Int. J. Parasitol. Drugs Drug Resist. 7, 262–271. https://doi.org/10.1016/j.ijpddr.2017.06.001.

Kuesel, A.C., 2016. Research for new drugs for elimination of onchocerciasis in Africa. Int. J. Parasitol. Drugs Drug Resist. 6, 272–286. https://doi.org/10.1016/j.ijpddr.2016.04.002.

Lacey, E., 1990. Mode of action of benzimidazoles. Parasitol. Today 6, 112–115. https://doi.org/10.1016/0169-4758(90)90227-U.

Lee, E.L., Iyngkaran, N., Grieve, A.W., Robinson, M.J., Dissanaike, A.S., 1976. Therapeutic evaluation of oxantel pamoate (1, 4, 5, 6-tetrahydro-1-methyl-2-[trans-3-hydroxystyryl] pyrimidine pamoate) in severe *Trichuris trichiura* infection. Am. J. Trop. Med. Hyg. 25, 563–567.

Lee, S.Y., Lim, J.K., 1978. A comparative study of the effect of oxantel-pyrantel suspension and mebendazole in mixed infections with *Ascaris* and *Trichuris*. Drugs 15 (Suppl 1), 94–98.

Levecke, B., Vercruysse, J., 2016. Pyrantel parasiticide therapy in humans. In: Alan, A. (Ed.), Pyrantel Parasiticide Therapy in Humans and Domestic Animals. Academic Press, pp. 109–128. https://doi.org/10.1016/B978-0-12-801449-3.00016-8.

Lim, J.K., 1978. Anthelminthic effect of oxantel and oxantel-pyrantel in intestinal nematode infections. Drugs 15 (Suppl 1), 99–103.

Lustigman, S., Prichard, R.K., Gazzinelli, A., Grant, W.N., Boatin, B.A., McCarthy, J.S., Basáñez, M.-G., 2012. A research agenda for helminth diseases of humans: the problem of helminthiases. PLoS Negl. Trop. Dis. 6. https://doi.org/10.1371/journal.pntd.0001582.

Lyons, E.T., Tolliver, S.C., 2012. Macrocyclic lactones for parasite control in equids. Curr. Pharmaceut. Biotechnol. 13, 1070–1077.

Marti, H., Haji, H.J., Savioli, L., Chwaya, H.M., Mgeni, A.F., Ameir, J.S., Hatz, C., 1996. A comparative trial of a single-dose ivermectin versus three days of albendazole for treatment of *Strongyloides stercoralis* and other soil-transmitted helminth infections in children. Am. J. Trop. Med. Hyg. 55, 477–481.

Martin, R.J., Clark, C.L., Trailovic, S.M., Robertson, A.P., 2004. Oxantel is an N-type (methyridine and nicotine) agonist not an L-type (levamisole and pyrantel) agonist: classification of cholinergic anthelmintics in *Ascaris*. Int. J. Parasitol. 34, 1083–1090. https://doi.org/10.1016/j.ijpara.2004.04.014.

McCarty, T.R., Turkeltaub, J.A., Hotez, P.J., 2014. Global progress towards eliminating gastrointestinal helminth infections. Curr. Opin. Gastroenterol. 30, 18–24. https://doi.org/10.1097/MOG.0000000000000025.

Moser, W., Ali, S.M., Ame, S.M., Speich, B., Puchkov, M., Huwyler, J., Albonico, M., Hattendorf, J., Keiser, J., 2016. Efficacy and safety of oxantel pamoate in school-aged children infected with *Trichuris trichiura* on Pemba Island, Tanzania: a parallel, randomised, controlled, dose-ranging study. Lancet Infect. Dis. 16, 53–60. https://doi.org/10.1016/S1473-3099(15)00271-6.

Moser, W., Schindler, C., Keiser, J., 2017a. Efficacy of recommended drugs against soil transmitted helminths: systematic review and network meta-analysis. BMJ 358, j4307. https://doi.org/10.1136/bmj.j4307.

Moser, W., Coulibaly, J.T., Ali, S.M., Ame, S.M., Amour, A.K., Yapi, R.B., Albonico, M., Puchkov, M., Huwyler, J., Hattendorf, J., Keiser, J., 2017b. Efficacy and safety of tribendimidine, tribendimidine plus ivermectin, tribendimidine plus oxantel pamoate, and albendazole plus oxantel pamoate against hookworm and concomitant soil-transmitted helminth infections in Tanzania and Côte d'Ivoire: a randomised, controlled, single-blinded, non-inferiority trial. Lancet Infect. Dis. 17, 1162–1171. https://doi.org/10.1016/S1473-3099(17)30487-5.

Moser, W., Sayasone, S., Xayavong, S., Bounheuang, B., Puchkov, M., Huwyler, J., Hattendorf, J., Keiser, J., 2018. Efficacy and tolerability of triple drug therapy with albendazole, pyrantel pamoate, and oxantel pamoate compared with albendazole plus oxantel pamoate, pyrantel pamoate plus oxantel pamoate, and mebendazole plus pyrantel pamoate and oxantel pamoate against hookworm infections in school-aged children in Laos: a randomised, single-blind trial. Lancet Infect. Dis. 18, 729–737. https://doi.org/10.1016/S1473-3099(18)30220-2.

Mrus, J., Baeten, B., Engelen, M., Silber, S.A., 2017. Efficacy of single-dose 500 mg mebendazole in soil-transmitted helminth infections: a review. J. Helminthol. 1–10. https://doi.org/10.1017/S0022149X17000426.

Naquira, C., Jimenez, G., Guerra, J., Bernal, R., Nalin, D., Neu, D., Aziz, M., 1989. Ivermectin for human strongyloidiasis and other intestinal helminths. Am. J. Trop. Med. Hyg. 40, 304–309.

Ndyomugyenyi, R., Kabatereine, N., Olsen, A., Magnussen, P., 2008. Efficacy of ivermectin and albendazole alone and in combination for treatment of soil-transmitted helminths in pregnancy and adverse events: a randomized open label controlled intervention trial in Masindi district, western Uganda. Am. J. Trop. Med. Hyg. 79, 856–863.

Nolan, T.J., Lok, J.B., 2012. Macrocyclic lactones in the treatment and control of parasitism in small companion animals. Curr. Pharmaceut. Biotechnol. 13, 1078–1094.

Olliaro, P., Seiler, J., Kuesel, A., Horton, J., Clark, J.N., Don, R., Keiser, J., 2011. Potential drug development candidates for human soil-transmitted helminthiases. PLoS Negl. Trop. Dis. 5, e1138. https://doi.org/10.1371/journal.pntd.0001138.

Opoku, N.O., Bakajika, D.K., Kanza, E.M., Howard, H., Mambandu, G.L., Nyathirombo, A., Nigo, M.M., Kasonia, K., Masembe, S.L., Mumbere, M., Kataliko, K., Larbelee, J.P., Kpawor, M., Bolay, K.M., Bolay, F., Attah, S.K., Vaillant, M., Halleux, C.M., Kuesel, A.C., 2018. Single dose moxidectin versus ivermectin for *Onchocerca volvulus* infection in Ghana, Liberia, and the Democratic Republic of the Congo: a randomised, controlled, double-blind phase 3 trial. in press. Lancet Lond. Engl. https://doi.org/10.1016/S0140-6736(17)32844-1.

Peldán, K., Pitkänen, T., 1982. Treatment of *Trichuris trichiura* infection with a single dose of oxantel pamoate. Scand. J. Infect. Dis. 14, 297–299.

Pullan, R.L., Smith, J.L., Jasrasaria, R., Brooker, S.J., 2014. Global numbers of infection and disease burden of soil transmitted helminth infections in 2010. Parasit. Vectors 7, 37.

Qian, H., Martin, R.J., Robertson, A.P., 2006. Pharmacology of N-, L-, and B-subtypes of nematode nAChR resolved at the single-channel level in *Ascaris suum*. FASEB J. Off. Publ. Fed. Am. Soc. Exp. Biol. 20, 2606–2608. https://doi.org/10.1096/fj.06-6264fje.

Qian, M.B., Yap, P., Yang, Y.C., Liang, H., Jiang, Z.H., Li, W., Utzinger, J., Zhou, X.N., Keiser, J., 2013. Accuracy of the Kato-Katz method and formalin-ether concentration technique for the diagnosis of *Clonorchis sinensis*, and implication for assessing drug efficacy. Parasit. Vectors 6, 314. https://doi.org/10.1186/1756-3305-6-314.

Rim, J.H., Lee, Y.M., Lee, J.S., Joo, K.H., 1981. Anthelmintic effects of single doses of fenbendazole and oxantel pyrantel pamoate to the intestinal nematodes. Kor. J. Parasitol. 19, 95–100.

Robertson, A.P., Puttachary, S., Buxton, S.K., Martin, R.J., 2015. Tribendimidine: mode of action and nAChR subtype selectivity in *Ascaris* and *Oesophagostomum*. PLoS Negl. Trop. Dis. 9. https://doi.org/10.1371/journal.pntd.0003495 e0003495.

Robertson, S.J., Pennington, A.J., Evans, A.M., Martin, R.J., 1994. The action of pyrantel as an agonist and an open channel blocker at acetylcholine receptors in isolated *Ascaris suum* muscle vesicles. Eur. J. Pharmacol. 271, 273–282.

Santos, A.R., dos Falcão, C.A.B., Muzitano, M.F., Kaiser, C.R., Rossi-Bergmann, B., Férézou, J.P., 2009. Ivermectin-derived leishmanicidal compounds. Bioorg. Med. Chem. 17, 496–502. https://doi.org/10.1016/j.bmc.2008.12.003.

Sayasone, S., Odermatt, P., Vonghachack, Y., Xayavong, S., Senggnam, K., Duthaler, U., Akkhavong, K., Hattendorf, J., Keiser, J., 2016. Efficacy and safety of tribendimidine against *Opisthorchis viverrini*: two randomised, parallel-group, single-blind, dose-ranging, phase 2 trials. Lancet Infect. Dis. 16, 1145–1153. https://doi.org/10.1016/S1473-3099(16)30198-0.

Sinniah, B., Sinniah, D., 1981. The anthelmintic effects of pyrantel pamoate, oxantel-pyrantel pamoate, levamisole and mebendazole in the treatment of intestinal nematodes. Ann. Trop. Med. Parasitol. 75, 315–321.

Sinniah, B., Sinniah, D., Dissanaike, A.S., 1980. Single dose treatment of intestinal nematodes with oxantel-pyrantel pamoate plus mebendazole. Ann. Trop. Med. Parasitol. 74, 619–623.

Soukhathammavong, P., Odermatt, P., Sayasone, S., Vonghachack, Y., Vounatsou, P., Hatz, C., Akkhavong, K., Keiser, J., 2011. Efficacy and safety of mefloquine, artesunate, mefloquine-artesunate, tribendimidine, and praziquantel in patients with *Opisthorchis viverrini*: a randomised, exploratory, open-label, phase 2 trial. Lancet Infect. Dis. 11, 110–118. https://doi.org/10.1016/S1473-3099(10)70250-4.

Speich, B., Ali, S.M., Ame, S.M., Bogoch, I.I., Alles, R., Huwyler, J., Albonico, M., Hattendorf, J., Utzinger, J., Keiser, J., 2015. Efficacy and safety of albendazole plus ivermectin, albendazole plus mebendazole, albendazole plus oxantel pamoate, and mebendazole alone against *Trichuris trichiura* and concomitant soil-transmitted helminth infections: a four-arm, randomised controlled trial. Lancet Infect. Dis. 15, 277–284. https://doi.org/10.1016/S1473-3099(14)71050-3.

Speich, B., Ame, S.M., Ali, S.M., Alles, R., Huwyler, J., Hattendorf, J., Utzinger, J., Albonico, M., Keiser, J., 2014. Oxantel pamoate–albendazole for *Trichuris trichiura* infection. N. Engl. J. Med. 370, 610–620. https://doi.org/10.1056/NEJMoa1301956.
Steinmann, P., Zhou, X.N., Du, Z.W., Jiang, J.Y., Xiao, S.H., Wu, Z.X., Zhou, H., Utzinger, J., 2008. Tribendimidine and albendazole for treating soil-transmitted helminths, *Strongyloides stercoralis* and *Taenia* spp.: open-label randomized trial. PLoS Negl. Trop. Dis. 2, e322. https://doi.org/10.1371/journal.pntd.0000322.
Uniting to Combat NTDs, 2012. London Declaration on Neglected Tropical Diseases. http://unitingtocombatntds.org/london-declaration-neglected-tropical-diseases/ (accessed January 9, 2018).
Vercruysse, J., Albonico, M., Behnke, J.M., Kotze, A.C., Prichard, R.K., McCarthy, J.S., Montresor, A., Levecke, B., 2011. Is anthelmintic resistance a concern for the control of human soil-transmitted helminths? Int. J. Parasitol. Drugs Drug Resist. 1, 14–27. https://doi.org/10.1016/j.ijpddr.2011.09.002.
Wen, L.Y., Yan, X.L., Sun, F.H., Fang, Y.Y., Yang, M.J., Lou, L.J., 2008. A randomized, double-blind, multicenter clinical trial on the efficacy of ivermectin against intestinal nematode infections in China. Acta Trop. 106, 190–194. https://doi.org/10.1016/j.actatropica.2008.03.007.
WHO, 2017a. Summary of global update on preventive chemotherapy implementation in 2016: crossing the billion - Wkly Epidemiol Rec, 92, 589–608.
WHO, 2017b. Preventive chemotherapy to control soil-transmitted helminth infections in at-risk population groups Guideline. Geneva World Health Organzation 1–75.
WHO, 2017c. WHO Model Lists of Essential Medicines - 20th List, Geneva World Health Organization.
WHO, 2017d. Guideline – Alternative mass drug administration regimens to eliminate lymphatic filariasis. Geneva World Health Organization.
WHO, 2015. WHO Model List of Essential Medicines - 19th List. Geneva World Health Organization.
WHO, 2013. Assessing the efficacy of anthelminthic drugs against schistosomiasis and soil-transmitted helminthiases. Geneva World Health Organization.
WHO, 2012. Soil-transmitted helminthiasis: eliminating soil-transmitted helminthiasis as a public health problem in children. Progress Report 2001-2010 and Strategic Plan 2011-2020. Geneva World Health Organization.
Wimmersberger, D., Coulibaly, J.T., Schulz, J., Puchkow, M., Huwyler, J., N'Gbesso, Y., Hattendorf, J., Keiser, J., 2018. Efficacy and safety of ivermectin against Trichuris trichiura in preschool- and school-aged children: a randomized controlled dose-finding trial. Clin. Infect. Dis. in press. https://doi.org/10.1093/cid/ciy246.
World Health Assembly, 2001. WHA54.19 Schistosomiasis and Soil-transmitted Helminth Infections.
Wu, Z., Fang, Y., Liu, Y., 2006. Effect of a novel drug–enteric coated tribendimidine in the treatment of intestinal nematode infections. Chinese journal of parasitology & parasitic diseases 24, 23–26.
Wu, Z., Qian, Y., 2003. Therapeutic effect of tribendimidine combined with ivermectin against human intestinal nematode infection. Chinese J Pract. Par. Diseases 2.
Xiao, S.H., Hui-Ming, W., Tanner, M., Utzinger, J., Chong, W., 2005. Tribendimidine: a promising, safe and broad-spectrum anthelmintic agent from China. Acta Trop. 94, 1–14. https://doi.org/10.1016/j.actatropica.2005.01.013.
Xiao, S.H., Utzinger, J., Tanner, M., Keiser, J., Xue, J., 2013. Advances with the Chinese anthelminthic drug tribendimidine in clinical trials and laboratory investigations. Acta Trop. 126, 115–126. https://doi.org/10.1016/j.actatropica.2013.01.009.

Xu, L.L., Jiang, B., Duan, J.H., Zhuang, S.F., Liu, Y.C., Zhu, S.Q., Zhang, L.P., Zhang, H.B., Xiao, S.H., Zhou, X.N., 2014. Efficacy and safety of praziquantel, tribendimidine and mebendazole in patients with co-infection of *Clonorchis sinensis* and other helminths. PLoS Negl. Trop. Dis. 8. https://doi.org/10.1371/journal.pntd.0003046 e3046.

Zhou, J.H., Lin, L., Zhang, Y.Y., 2008. Therapeutic effect of tribendimidine in treatment of *Enterobius vermicularis* infection. Chin. J. Schistosomiasis Control 226.

CHAPTER SIX

Community and Drug Distributor Perceptions and Experiences of Mass Drug Administration for the Elimination of Lymphatic Filariasis: A Rapid Review of Qualitative Research

Heather Melanie R. Ames[*,§,¶,1], **Meike Zuske**[§,¶], **Jonathan D. King**[||], **Peter Steinmann**[§,¶] **and Xavier Bosch-Capblanch**[§,¶,1]
[*]Norwegian Institute of Public Health, Division for Health Services, Oslo, Norway
[§]Swiss Tropical and Public Health Institute, Basel, Switzerland
[¶]University of Basel, Basel, Switzerland
[||]Department of Control of Neglected Tropical Diseases, World Health Organisation, Geneva, Switzerland
[1]Corresponding authors: E-mails: heather.ames@fhi.no (HA); x.bosch@unibas.ch (XBC).

Contents

Advances in Parasitology, Volume 103
ISSN 0065-308X
https://doi.org/10.1016/bs.apar.2018.09.003

Abstract

Study objectives:

This article presents findings from a rapid review of qualitative research conducted to inform decision makers about community and drug distributor perceptions and experiences of mass drug administration campaigns for the elimination of lymphatic filariasis. We focused on questions related to acceptability of the mass drug administration campaigns within these groups and their thoughts around the feasibility of planning and carrying out the campaigns.

Methods:

We carried out a systematic search in five databases to identify potential studies. We included studies that focused on community members and drug distributors and used qualitative methods for data collection and analysis. We conducted a thematic framework analysis using the Supporting the Use of Research Evidence framework. Due to time constraints, one author conducted the screening, extraction and data analysis.

Findings:

Studies found that communities lack knowledge and information about lymphatic filariasis and the mass drug administration campaigns and that this can have an impact on how many community members choose to take medication. Health workers often had a good understanding that lymphatic filariasis was a problem in their setting, of its cause and mode of transmission and that hydrocele and elephantiasis had the same cause. However, this knowledge was not as prevalent amongst community drug distributors who often had misconceptions surrounding the topic. Furthermore, studies found that the length, timing, level of community and health system involvement, access to care for side effects, inadequate numbers of drug distributors and supervisors and motivation of drug distributors influenced participation in mass drug administration campaigns. Finally, the inadequate training of drug distributors could influence community trust in the mass drug administration program and the drug distributor themselves if there was a perception that the person was uninformed or not trained to carry out their tasks.

1. INTRODUCTION

Lymphatic filariasis (LF) is a vector-borne neglected tropical disease caused by *Wuchereria bancrofti* as well as *Brugia malayi* and *B. timori*, the latter two parasites with restricted endemicity in Asia. Transmission occurs when microfilariae, released by female adult worms and circulating in the blood, are ingested by mosquitoes, become infectious larva and are spread to other hosts as mosquitoes continue to feed. Multiple species of *Aedes*, *Anopheles*, *Culex*, and *Mansonia* have been documented as vectors of lymphatic filariasis. The adult parasites nest in the lymphatic vessels leading to obstruction and impaired lymphatic function. Hydrocele and lymphoedema (elephantiasis) are the well-known disabling manifestations.

In 2000, WHO launched The Global Programme to Eliminate Lymphatic Filariasis (GPELF) which aims by 2020 to stop transmission of the infection and alleviate suffering among those affected with the disease (Ichimori et al., 2014). At that time, around 120 million persons were estimated infected with approximately 40 million persons living with lymphedema or hydrocele (Michael et al., 1996). Estimates incorporating the impact of GPELF through 2013 suggest a reduced burden of 68 million cases with 36 million persons with chronic manifestations (Reuters, 2016). The estimated Disability Adjusted Life Years due to lymphatic filariasis has decreased from 1.9 million in 2000 to 1.2 million in 2016 (WHO, 2016). As of 2016, WHO considered 72 countries to be endemic and a population of 856 million persons living in areas of 52 countries which require mass drug administration (MDA) to stop transmission of infection (WHO, 2017a).

MDA reduces circulating microfilariae in the blood of infected persons and prevalence in the community. Several rounds of MDA are required to reduce infection below levels at which transmission can no longer be sustained. Effectiveness of MDA depends in part on the proportion of persons in the community ingesting the medicines. Therefore, MDA campaigns aim to reach all eligible individuals in endemic districts. At least 65% of the total population ingesting the medicine is referred to as 'effective coverage' and considered as the minimum coverage required having the desired impact on transmission. WHO recommends at least five annual rounds of MDA with at least 65% coverage of the total population before evaluating impact on infection levels (WHO, 2011).

A new MDA regimen of ivermectin, diethylcarbamazine and albendazole (IDA) is more effective at clearing microfilariae from the blood than the current two-drug regimens, is felt to reduce transmission to below elimination thresholds in fewer MDA rounds and is now recommended for MDA in special settings (WHO, 2017b). Drug donations have facilitated access to all three medicines recommended for MDA. The availability of effective drugs and treatment regimens are essential but alone are not sufficient to eliminate LF (Kastner et al., 2015). Mathematical models clearly predict that even with a more efficacious MDA regimen, gains will not be observed unless compliance is maximised (Irvine et al., 2017). Additionally, MDA requires complex coordination between and within countries and health systems for delivery, reporting and ultimately acceptance by targeted communities (Gyapong et al., 2018).

With the promise of potentially more effective, alternative treatment regimens, we rapidly reviewed the existing qualitative literature to identify perspectives from the community and drug distributors. We focused on factors influencing feasibility of planning and carrying out campaigns and acceptability of MDA within community settings with the goal of informing decision makers about community and drug distributor perceptions and experiences of MDA campaigns for the elimination of LF.

2. METHODS

This is a rapid review of qualitative primary studies to explore and present a summary of the data available on the topic (Haby et al., 2016). Due to time constraints, one author conducted the screening, extraction and analysis and a GRADE CERQual assessment was not conducted (Lewin et al., 2015).

Our research focused on community members and drug distributor's perceptions of and experiences with MDA campaigns for the elimination of LF, regardless of the specific treatment. We define acceptability as the extent to which that intervention is considered to be reasonable among those receiving, delivering or affected by the intervention (WHO, 2014), Community participation, taking the medicines, is an indicator for acceptability among those receiving the intervention. The feasibility of an intervention or option being considered in a guideline is the likelihood that it can be properly carried out or implemented in a given context (WHO, 2014). Feasibility is influenced by the nature of the intervention itself; the human and other material resources required to deliver it; the recipients of the intervention and other stakeholders; the characteristics of the health system; and social, political and other contextual factors (SURE Collaboration, 2011).

We included all studies that used qualitative methods for data collection (e.g. focus group interviews, individual interviews, observation, document analysis) and that utilised qualitative methods for data analysis (e.g. thematic analysis, framework analysis, grounded theory) that discussed community and/or drug distributor perceptions of and experiences with any form of MDA for LF elimination. We excluded studies that collected data using qualitative methods but did not perform a qualitative analysis (e.g. open-ended survey questions where the responses are analysed using descriptive statistics). We included mixed-methods studies where it was possible to extract findings derived from qualitative research. However, for a number of these studies extracted data may have come from a combination of survey and qualitative data, as authors did not always distinguish the source of the findings specifically in the published article.

We included studies that focused on community members and drug distributors. Community encompasses people receiving treatment as well as those around them. A drug distributor can be anyone distributing medicines.

2.1 Search Methods for Identification of Studies

We searched the following electronic databases for eligible studies published between 2002 and February 1st, 2017: Global Health Library, Embase (Ovid), MEDLINE (OvidSP), SCOPUS, Web of Science.

The search strategy was developed by an information specialist using guidelines developed by the Cochrane Qualitative Research Methods Group for searching for qualitative evidence (Noyes et al., 2008) as well as pulling keywords and MeSH terms from a search for a qualitative evidence synthesis with a similar scope (Ames et al., 2017b). There was no language or geographic restrictions for the search. The complete search strategy for this review can be found in the supplementary material. Additionally, we conducted citation searches of included studies in Google Scholar, Scopus and Web of Science and asked experts of the field to send in relevant study titles.

We collected data by compiling all records identified into one Endnote (Reuters, 2016) database. After duplicated records were removed, each record was individually assessed for relevance by the first author who then retrieved the full text of all relevant papers and reviewed them for inclusion.

The protocol of this study was registered in PROSPERO and is available at https://www.crd.york.ac.uk/PROSPERO/display_record.asp?ID=CRD42017057253.

3. DATA COLLECTION AND ANALYSIS

The first author performed data extraction using a data extraction form designed specifically for this synthesis. The form was used to extract key themes and categories relevant to the synthesis objective as well as information about first author, date of publication, language, country of study, context (e.g. urban, rural), participant group (e.g. infected person, relative, community leader), theoretical or conceptual framework, and research methods.

The first author assessed the methodological quality of the included studies by applying an adaption of the Critical Appraisal Skills Programme (CASP) (CASP, 2017) which has been used in other qualitative evidence syntheses (Glenton et al., 2013b; Carlsen et al., 2007; Munro et al., 2007; Ames et al., 2017b). We accept that there is no "gold standard" approach for assessing the methodological quality of primary qualitative studies, but believe that this adapted CASP checklist best fit our needs.

A thematic framework analysis was conducted using the Supporting the Use of Research Evidence (SURE) framework (SURE Collaboration, 2011). The framework has been used as an analysis framework in other studies and reviews (Gopinathan et al., 2014; Muloliwa AM et al., 2017; Oku A et al., 2017; Glenton et al., 2013a; Glenton et al., 2016; Lewin et al., 2010). The headings and sub headings from the SURE framework were used to construct the key messages presented below. Findings were then regrouped based on similarities and themes.

4. ROLE OF THE FUNDING SOURCE

This review was part of a set of systematic reviews funded by the World Health Organisation (WHO) to develop a global public health guideline on alternative MDA regimens for LF elimination (WHO, 2017b). One of the article's authors is working at the Department of Control of Neglected Tropical Diseases of the WHO and was the author on one of the studies included in the synthesis (King et al., 2011). The funder had no role in the inclusion or exclusion of any study from the analysis.

5. RESULTS

Our database search yielded 691 references. An additional reference was obtained from an expert. After removing 207 duplicates, 485 references

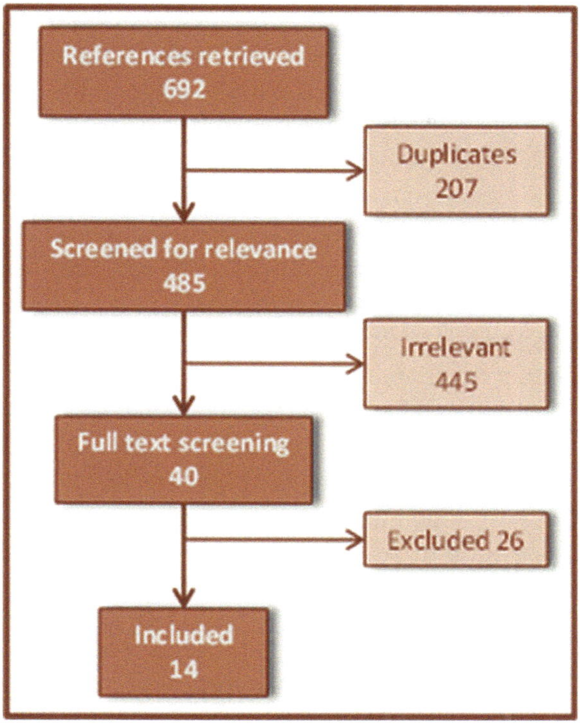

Figure 1 Flow of studies.

were screened for relevance by the first author. Forty references were identified as relevant and retrieved in full text. After applying the inclusion criteria, and when appropriate contacting study authors, 14 studies were included in the analysis (Fig. 1).

A description of the included studies can be found in Appendix 1.

5.1 Characteristics of Included Studies

The fourteen included studies were published between 2004 and 2016 and took place in low and middle-income countries (The World Bank Data; Coun, 2016): Philippines (Amarillo et al., 2008), Dominican Republic (Baker et al., 2007), Ghana (da-Costa Vroom et al., 2015), American Samoa (King et al., 2011), Papua New Guinea (Wynd et al., 2007), Indonesia (Krentel and Aunger, 2011), Tanzania (Kisoka et al., 2016a, 2016b), Kenya (Njomo et al., 2012, 2014), and India (Hussain et al., 2014; Babu, 2010; Babu and Kar, 2004; Babu and Nath, 2004).

Five of the included studies were mixed methods studies that employed some qualitative research methods and reported the qualitative and quantitative findings in the same article (Amarillo et al., 2008; Babu and Kar, 2004; Baker et al., 2007; Hussain et al., 2014; King et al., 2011). Two studies used qualitative data collection methods and analysed the data qualitatively based on frequency of participants' responses (Babu, 2010; Babu and Nath, 2004). Four studies reported the qualitative findings of the qualitative portion of a larger mixed methods study (Kisoka et al., 2016a; Njomo et al., 2012, 2014; Krentel and Aunger, 2011). Only three of the included studies were purely qualitative research based on the descriptions within the studies themselves (da-Costa Vroom et al., 2015; Kisoka et al., 2016b; Wynd et al., 2007).

Health workers or community drug distributors (CDDs) were the sole participants in the qualitative methods in two studies (Baker et al., 2007; Hussain et al., 2014). Community members were the sole participants in the qualitative methods in two studies (Wynd et al., 2007; Krentel and Aunger, 2011). The remaining studies included both community participants and health workers/CDDs in the qualitative portion of the research.

5.2 Key Messages of the Rapid Review

Summary of key message tables are presented in appendix 2 and detail the included primary studies, study setting, and methodological assessment contributing to each key message. An overview of key messages is presented in Table 1.

5.3 Community Members Knowledge Regarding LF

Key message 1: Studies found that in general communities lacked knowledge about the cause and mode of transmission of LF and that hydrocele and elephantiasis had the same cause.

Some studies (Amarillo et al., 2008; Babu and Nath, 2004; Kisoka et al., 2016a, 2016b; Wynd et al., 2007; Krentel and Aunger, 2011) discussed community members' knowledge of LF and the MDA campaigns. In Tanzania, neglected tropical diseases were not well known among communities in endemic areas (Kisoka et al., 2016a). In Tanzania (Kisoka et al., 2016b) and Papua New Guinea (Wynd et al., 2007), most community members did not view LF as a current health problem. Some participants in Indonesia (Krentel and Aunger, 2011) considered if LF was contracted, it would detract from their economic earnings.

Table 1 Key Messages From the Qualitative Rapid Review

Key Message		Number of Included Studies
Community Members Knowledge Regarding LF		
1	Studies found that in general communities lacked knowledge about the cause and mode of transmission and that hydrocele and elephantiasis had the same cause.	6
Attitudes Towards Programme Acceptability		
2	Studies found that when communities understood the benefits and aims of MDA campaigns for the elimination of LF and the potential side effects of the tablets, they were more open to and accepting of the MDA campaigns.	8
3	Studies found that in many instances it was important for the community member to know the person distributing the drugs. In some instances, community members preferred health workers distribute the drugs as they believed they were more competent to conduct the task due to their medical training.	7
4	Studies found that in general, door-to-door distribution was accepted and that some communities preferred distribution at religious events and services.	5
5	Studies found that the planning of MDA campaigns should take local community events such as harvest, other public health campaigns and religious or national holidays into account. Consultation with the community and a timely implementation of the campaigns were considered key to influencing community members' attitudes towards the MDA campaigns.	5
6	Studies found that communities wanted more frequent messaging about MDA campaigns delivered through a wider variety of locally preferred communication modes.	7
7	Studies found that communities felt that they were not receiving enough information about MDA campaigns. They wanted more information about the benefits of and rationale for participating and the side effects from taking the tablets.	10
Motivation to Participate in the MDA Campaigns to Eliminate LF and Take the Tablets Distributed		
8	Studies found that participation in MDA campaigns tended to be higher when; communities are involved in the planning and implementation of the campaign, the disease (LF) and MDA campaigns are considered a priority in the community, and those distributing the drugs, whether health workers or CDDs, are trusted in the community setting. This community involvement in the planning and implementation of MDA campaigns was important in ensuring their success.	8

(Continued)

Table 1 Key Messages From the Qualitative Rapid Review—cont'd

Key Message		Number of Included Studies
9	Studies found that the involvement of and support from all levels of healthcare services in the MDA campaigns was important for the success of the programme.	9
10	Studies found that participation in MDA campaigns tends to decrease when; side effects are experienced or feared, when there is a lack of information about LF, the tablets or the campaign available to community members or when communities believe rumours about the tablets or campaigns that raise doubts and suspicion.	6
11	Studies found that access to and the provision of treatment for side effects during MDA campaigns could help to increase coverage and compliance.	5
Health Worker and Drug Distributor Knowledge of LF and MDA Campaigns		
12	Studies found that often health workers had a good understanding that LF was a problem in their setting, of the cause and mode of transmission of LF and that hydrocele and elephantiasis had the same cause. However, this knowledge was not as prevalent amongst community drug distributors who often had misconceptions surrounding the topic.	5
Recruitment and Training of Drug Distributors		
13	The recruitment and selection of drug distributors varied across contexts. However, the majority of participants believed that the selection process should be transparent and the selection criteria clearly described in order to avoid conflict.	3
14	Studies found that in general community drug distributors did not receive adequate training to fulfil their duties during MDA campaigns.	7
Health Worker and CDD Attitudes Towards Program Acceptability		
15	Health workers and drug distributors often had positive attitudes towards the MDA campaigns viewing them as useful and reducing the problems associated with LF in their communities. Health workers and drug distributors had more negative attitudes towards the drug distribution process.	5
Health Worker and CDD Motivation to Participate in the MDA Campaigns		
16	Health workers and CDDs were motivated to participate in MDA campaigns when they saw a decrease in the number of LF cases in their communities, an increase in community awareness and knowledge of LF, were satisfied with their jobs and felt pride in	7

Table 1 Key Messages From the Qualitative Rapid Review—cont'd

	Key Message	Number of Included Studies
	their work. Difficult physical conditions, a lack of incentives, insufficient staffing of distributors and insufficient planning and implementation of awareness activities around campaigns were demotivating.	
17	Studies found that incentives were important for drug distributor motivation when they were perceived as adequate and a demonstration of the appreciation of their time and effort. However, if the remuneration, financial or other, was perceived as not enough or less than what was offered in other public health programmes, it influenced distributor motivation and could lead to leaving the programme.	5
18	Studies found that the majority of health workers and community drugs distributors believed that the MDA campaigns were too short and did not provide enough time to reach every household. In some cases, it was reported that there was an insufficient number of people involved in distributing tablets during MDA campaigns.	8
19	Studies found that there was a lack of supervision of community drug distributors during MDA campaigns.	6

MDA campaigns in the Philippines led to an increase in knowledge and awareness of LF and helped to clarify community misconceptions (Amarillo et al., 2008). In contrast, community misperceptions persisted even after MDA campaigns in Papua New Guinea (Wynd et al., 2007). In Tanzania (Kisoka et al., 2016b), elephantiasis and hydrocele were assigned different causes and there was confusion over the cause and mode of transmission of LF even after education during MDA campaigns. In India, most community members knew that LF could be eliminated. However, few participants knew the benefits of MDA or that LF could be prevented by taking the medications (Babu and Nath, 2004).

5.4 Attitudes Towards Programme Acceptability

Key message 2: Studies found that when communities understood the benefits and aims of MDA campaigns for the elimination of LF and the potential side effects of the tablets, they were more open to and accepting of the MDA campaigns.

A few studies (Wynd et al., 2007; Babu and Kar, 2004; Kisoka et al., 2016b; Krentel and Aunger, 2011) found that when communities understood the benefits and aims of the MDA campaigns they were more open and accepting of them. They trusted that MDA was beneficial to individuals and the community for disease prevention, cure and rehabilitation (Kisoka et al., 2016b; Wynd et al., 2007; Krentel and Aunger, 2011). They understood that the goal of the programme was disease elimination (Kisoka et al., 2016b; Krentel and Aunger, 2011).

Some community members believed in the positive effects of the drugs themselves (Wynd et al., 2007; Kisoka et al., 2016a, 2016b; Krentel and Aunger, 2011). They reported benefiting from the drugs when symptoms such as itching disappeared after MDA campaigns (Kisoka et al., 2016a, 2016b; Wynd et al., 2007). Some believed that the tablets also cured other sicknesses (Wynd et al., 2007). Some people viewed the MDA campaigns more favourably when they were made aware of potential side effects before the drugs were distributed (Babu and Kar, 2004).

Many studies (Kisoka et al., 2016a, 2016b; Babu, 2010; Babu and Kar, 2004; Babu and Nath, 2004; Njomo et al., 2012; Krentel and Aunger, 2011) presented findings related to negative community perceptions of MDA campaigns. The MDA campaign was not a priority for residents living in an area that lacked basic health care services. The community felt that the government should be focussing on more important aspects of health care than the elimination campaigns for LF (Kisoka et al., 2016b).

In some studies (Kisoka et al., 2016b; Babu and Kar, 2004; Babu and Nath, 2004; Krentel and Aunger, 2011) the MDA campaign raised suspicions amongst community members for various reasons; for example, unclear programme motives (Babu and Kar, 2004), rumours of serious side effects including sterilisation and death (Babu and Kar, 2004; Babu, 2010; Babu and Nath, 2004; Kisoka et al., 2016b; Krentel and Aunger, 2011), the drugs were "not good" (Babu and Nath, 2004), or the fact that the drugs were free (Kisoka et al., 2016b).

Key message 3: Studies found that in many instances it was important for the community member to know the person distributing the drugs. In some instances, community members preferred health workers distribute the drugs, as they believed they were more competent to conduct the task due to their medical training.

Studies from Kenya (Njomo et al., 2012, 2014) and Tanzania (Kisoka et al., 2016a, 2016b) discussed findings related to those distributing medications during MDA campaigns. In both settings, distributors spoke of their sometimes negative and sometimes positive reception in the community. It was important to community members that the distributor was someone they knew and not a stranger (Kisoka et al., 2016a; Njomo et al., 2012). In Tanzania, some distributors and community members believed that health workers should distribute the drugs. Some refused to take the tablets when they felt that the non-health workers distributing the drugs were not properly trained (Kisoka et al., 2016a, 2016b). A few studies (King et al., 2011; Wynd et al., 2007; Krentel and Aunger, 2011) found that communities were more accepting of MDA campaigns when community organisations, community leaders and churches were involved in drug distribution and seen to be publicly supporting disease elimination efforts.

Key message 4: Studies found in general, door-to-door distribution was accepted and that some communities preferred distribution at religious events and services.

Some studies (King et al., 2011; Kisoka et al., 2016a; Njomo et al., 2014; Wynd et al., 2007; Krentel and Aunger, 2011) presented findings linked to the perceptions of the location in which drugs were distributed. In most contexts, drugs were distributed in public settings and door-to-door distribution. In Kenya (Njomo et al., 2014), participants were satisfied with door-to-door distribution but felt that not every house was reached due to the short duration of the campaign. Door-to-door distribution allowed families to take the medications together, people could be reached in their own homes and they felt that the distributors would ensure that people who should not be taking the drugs did not receive them.

In American Samoa (King et al., 2011) door-to-door distribution was discontinued in 2003 and a focus on distribution in locations where people naturally gathered such as schools, place of employment, churches and the airport became the focus. Nurses interviewed, preferred the house-to-house distribution as it provided a more personal interaction with people. However, they understood the logistical and time limitations that went with that form of distribution.

In American Samoa (King et al., 2011) and Papua New Guinea (Wynd et al., 2007) there was a preference for drug distribution to follow religious services as this was when most of the community members were gathered together. In Indonesia (Krentel and Aunger, 2011), participants felt that when distribution occurred in public settings, such as after a religious service, there was increased public pressure to comply as you did not want to be seen as the only one not swallowing the tablets.

Key message 5: Studies found that the planning of MDA campaigns should take local community events such as harvest, other public health campaigns and religious or national holidays into account. Consultation with the community and a timely implementation of the campaigns were considered key to influencing community members' attitudes towards the MDA campaigns.

Some studies (Babu and Kar, 2004; da-Costa Vroom et al., 2015; Kisoka et al., 2016a; Kisoka et al., 2016b) discussed findings related to the timing of the MDA campaigns. If the timing of the drug distribution campaign was inconvenient to residents, for example during harvest or Ramadan, or conflicted with another public health campaign, they perceived the campaign negatively (Kisoka et al., 2016a; da-Costa Vroom et al., 2015).

Findings related to the financial resources and the impact on campaign implementation were mentioned by a few studies (da-Costa Vroom et al., 2015; Njomo et al., 2012; Kisoka et al., 2016a). When campaigns failed to initiate on the scheduled day and were delayed, often due to delayed release of funds or medication (Kisoka et al., 2016a), compliance also decreased. There were a number of factors contributing to this decrease including; the advertising in newspapers had already occurred and the community was not informed of a delay (Babu and Kar, 2004), the delayed release of education and communication materials (Kisoka et al., 2016b), difficulty recruiting distributors, as they preferred to take the jobs that paid better (da-Costa Vroom et al., 2015), or led the campaigns to coincide with the rainy seasons (when farmers are away) or Ramadan (where drugs can only be consumed after dark) (Kisoka et al., 2016a).

Key message 6: Studies found that communities wanted more frequent messaging about MDA campaigns delivered through a wider variety of locally preferred communication modes.

A few studies mentioned findings related to the frequency of communicating about MDA campaigns (Kisoka et al., 2016a; Njomo et al., 2014). They found that participants wanted more frequent messaging to occur in the run up to MDA campaigns so that communities would be more aware when distributors arrived at their homes.

Health workers play an important role in communicating about LF and MDA to communities in some settings. In the Philippines (Amarillo et al., 2008), health workers were found to be the major source of information about LF and MDA for community members. However, health workers felt like they needed more support from local government to make the programme succeed. In Tanzania (Kisoka et al., 2016a), it was the health workers responsibility to reach out to communities to discuss LF and underline the importance of collaboration between health workers and communities during MDA campaigns.

Findings related to IEC strategies were also presented (Amarillo et al., 2008; Babu and Kar, 2004; Njomo et al., 2014). In the Philippines, NGO staff were very involved in IEC activities but knew little of the rationale or benefits of the MDA (Amarillo et al., 2008). In India, the authors discovered that in communities where IEC activities had been better, communities understood the benefits of the programme and were more aware of the side effects. In these communities, coverage and compliance rates tended to be higher (Babu and Kar, 2004). A more detailed account of IEC activities in Kenya was presented (Njomo et al., 2014). CDDs wanted IEC materials available in the local language with frequent reminders by town announcers in the days before the campaign to ensure community awareness. They felt that the posters in use had adequate information and a good use of visuals but that more detail was needed surrounding the cause of the disease. Opinion leaders in this setting expressed the need for a combination of IEC interventions, such as posters, town announcers, banners and artistic performances, to increase community awareness.

A lack of information about the campaign itself or the medications included was found to be a reason for non-participation in MDA campaigns (Babu and Kar, 2004; Kisoka et al., 2016a, 2016b; Njomo et al., 2012, 2014; Krentel and Aunger, 2011). Villagers in India did not know the date or time for when the medications were going to be distributed (Babu and Kar, 2004). In Tanzania (Kisoka et al., 2016a), participants described being surprised and intimidated when drug distributors showed up at their houses as they had not received any information about the campaign and did not know it was happening.

Community members in Kenya were not satisfied with their communication and interaction with the CDDs. They believed that they had poor communication skills and had not given adequate information about the drugs (Njomo et al., 2012).

"But the problem is that people were not educated on the drugs and the CDDs just came and gave out the drugs, they did not explain anything that is why many people did not swallow the drugs." (Male youth FGD participant in a high compliance village) (Njomo et al., 2012).

Key message 7: Studies found that communities felt that they were not receiving enough information about MDA campaigns. They wanted more information about the benefits of and rationale for participating and the side effects from taking the tablets.

Many studies (Amarillo et al., 2008; Babu, 2010; Babu and Kar, 2004; Kisoka et al., 2016a; Njomo et al., 2014; Wynd et al., 2007; King et al., 2011) referred to mass media as a communication tool for MDA campaigns. Television and radio public service announcements, ranging from educational videos to announcements of date and time, were used to inform communities about upcoming campaigns (Amarillo et al., 2008; Babu and Kar, 2004; Kisoka et al., 2016a; Njomo et al., 2014; Wynd et al., 2007; King et al., 2011). In India (Babu, 2010; Babu and Kar, 2004), mass media had a negative effect on uptake when local newspapers reported rumours of exaggerated side effects, such as death, from the MDA campaigns.

A number of studies addressed the lack of information or lack of wanted content within the information that was received (Babu, 2010; Babu and Kar, 2004; Kisoka et al., 2016a, 2016b; Njomo et al., 2012, 2014; Amarillo et al., 2008; Krentel and Aunger, 2011). The findings highlight that community members wanted more information about side effects (Babu, 2010) and the benefits and rationale for taking the drugs (Kisoka et al., 2016a). One study highlighted that this kind of information was not provided in local languages (Amarillo et al., 2008).

5.5 Motivation to Participate in the MDA Campaigns to Eliminate LF and to Take the Tablets Distributed

Table 2 provides an overview of factors influencing community members' choice to participate in MDA campaigns.

Table 2 Reasons Influencing Community Participation in MDA Campaigns

Community Reasons to Participate in MDA	Community Reasons for not Participating in MDA
- High levels of community involvement in planning an implementation of drug distribution (Babu and Kar, 2004) - The tablets were from the government so they must be accepted (Krentel and Aunger, 2011)	- The occurrence of side effects (Amarillo et al., 2008; Babu and Nath, 2004; Babu and Kar, 2004; Babu, 2010; Wynd et al., 2007; Krentel and Aunger, 2011) - The impact side effects would have on their income (i.e. having to take a day off work) (Krentel and Aunger, 2011) (Amarillo et al., 2008)

Table 2 Reasons Influencing Community Participation in MDA Campaigns—cont'd

Community Reasons to Participate in MDA	Community Reasons for not Participating in MDA
- Social pressure to conform especially when tablets were distributed in public areas (Krentel and Aunger, 2011) - Fear of others not complying (Krentel and Aunger, 2011) - Having become a community leader and wanting to set a good example (Krentel and Aunger, 2011) - To maintain a good social reputation (Krentel and Aunger, 2011) - If LF is considered a priority than the programme was considered beneficial (Babu and Kar, 2004) (Wynd et al., 2007) - Believing the advice of a trusted health worker (Amarillo et al., 2008)	- Lack of information about the campaign and the included medications (i.e. when the medication would be distributed or that a campaign was happening) (Babu and Kar, 2004; Kisoka et al., 2016a, 2016b; Njomo et al., 2012, 2014; Krentel and Aunger, 2011) - Rumours that the drugs being distributed were for sterilisation (Njomo et al., 2012) (Kisoka et al., 2016b) - Reports of fatalities due to the drugs (Babu, 2010; Babu and Kar, 2004; Babu and Nath, 2004; Krentel and Aunger, 2011) - Lack of trust (Babu and Kar, 2004) - Loose tablets without labels arouse suspicion (Babu and Kar, 2004) - An excessive number of pills (Wynd et al., 2007) - Forgetfulness (Wynd et al., 2007; Krentel and Aunger, 2011) - Too busy (Krentel and Aunger, 2011) - Had heard that everyone else had complied so did not feel the need to take the tablets as the community was protected (Krentel and Aunger, 2011) - Having to take time off work to go and swallow the tablets (Krentel and Aunger, 2011) - Contraindications such as being pregnant or breastfeeding (Wynd et al., 2007; Krentel and Aunger, 2011) - That the drugs were chemically manufactured (not natural) and participants were unsure of what they would do to their bodies (Krentel and Aunger, 2011) - Had tested negative for LF (Krentel and Aunger, 2011)

Key message 8: Studies found that participation in MDA campaigns tended to be higher when; communities were involved in the planning and implementation of the campaign, the disease (LF) and MDA campaigns are considered a priority in the community, and those distributing the drugs, whether health workers or CDDs, are trusted in the community setting. This community involvement in the planning and implementation of MDA campaigns was important in ensuring their success.

A few articles (Wynd et al., 2007; Babu and Kar, 2004; Krentel and Aunger, 2011) presented information about why communities participated in MDA campaigns. The authors highlighted two factors. The first was the positive impact of community involvement. A number of studies discussed community participation in various facets of MDA campaigns (Babu and Kar, 2004; Baker et al., 2007; King et al., 2011; Kisoka et al., 2016a; Njomo et al., 2012, 2014; Wynd et al., 2007). All of the studies found that community involvement was a key element in their success. Two studies reported limited or inadequate community involvement as a hindrance to MDA campaigns (Babu and Kar, 2004; Njomo et al., 2012).

In American Samoa (King et al., 2011), Kenya (Njomo et al., 2014) and Papua New Guinea (Wynd et al., 2007) the preferred method of community involvement was through religious organisations. In Kenya (Njomo et al., 2014) and Papua New Guinea (Wynd et al., 2007) the community thought that the religious groups could be better utilised to communicate about MDA campaigns and aid in distribution. In Tanzania (Kisoka et al., 2016a), the central focus of community involvement was through social mobilisation. In American Samoa, involvement of other community settings such as schools, work places, bingo halls, health stations and airports were mentioned as alternative strategies to explore for community participation (King et al., 2011).

The second factor contributing to increased participation was whether LF and the MDA campaign were considered a priority. In communities where LF was considered a priority and the programme was considered beneficial, coverage rates were higher (Babu and Kar, 2004; Wynd et al., 2007).

Key message 9: Studies found that the involvement of and support from all levels of healthcare services in the MDA campaigns was important for the success of the programme.

A number of studies (Njomo et al., 2012; Njomo et al., 2014; Amarillo et al., 2008; Babu, 2010; Babu and Kar, 2004; Baker et al., 2007; da-Costa Vroom et al., 2015; Hussain et al., 2014; Kisoka et al., 2016b) mentioned findings related to human resources use during MDA campaigns. The most frequently mentioned was the importance of the involvement and support of all levels of health care services in the MDA campaigns (Amarillo et al., 2008; Babu and Kar, 2004; Baker et al., 2007; Njomo et al., 2014). In many instances studies reported that when this involvement and support existed compliance increased (Amarillo et al., 2008; Babu and Kar, 2004; Baker et al., 2007). In other cases increased support from the higher levels of the health services and government were needed to increase the success of the programme (Amarillo et al., 2008; Baker et al., 2007).

Key message 10: Studies found that participation in MDA campaigns tends to decrease when; side effects are experienced or feared, when there is a lack of information about LF, the tablets or the campaign available to community members or when communities believe rumours about the tablets or campaigns that raise doubts and suspicion.

Participants in some studies (Amarillo et al., 2008; Babu and Nath, 2004; Babu and Kar, 2004; Babu, 2010; Wynd et al., 2007; Krentel and Aunger, 2011) made a direct link between the side effects they or others had experienced from the medications and the decision not to take the medications again. In Indonesia (Krentel and Aunger, 2011) and the Philippines (Amarillo et al., 2008) community members decided to avoid taking the drugs due to the impact that the side effects would have on their income. Mostly day wage labourers felt that they could not risk losing a day's work due to adverse reactions to the LF medication. A fear of side effects in the Indian context lead to subsequent mop up activities being cancelled in many areas (Babu, 2010; Babu and Kar, 2004).

Key message 11: Studies found that access to and the provision of treatment for side effects during MDA campaigns could help to increase coverage and compliance.

In some cases, when general health care and services were poor and or lacking, it lead to a decreased trust in health workers and communities to believe that the MDA campaigns were not a good use of resources (Kisoka et al., 2016a, 2016b). Three articles (Babu and Kar, 2004; Babu, 2010; Njomo et al., 2012) raised findings related to the accessibility to treatment for side effects after the tablet ingestion. In India it was reported that local primary health centres managed side effects (Babu and Kar, 2004), CDDs were able to distribute medication for side effects (Babu, 2010), and doctors from mobile health units were on hand to help deal with adverse reactions (Babu, 2010). In this setting, side effects were not perceived to be a big problem for the MDA campaigns by health workers. However, in contrast, community members believed that they were a large problem and one of the main reasons for not taking the tablets (Babu, 2010). In areas where side effects were treated by health personnel compliance seemed to be higher (Babu and Kar, 2004). In Kenya a number of community members highlighted the importance of side effect control to increase compliance levels (Njomo et al., 2012).

5.6 Health Worker and Drug Distributor Knowledge of LF and MDA Campaigns

Key message 12: Studies found that often health workers had a good understanding that LF was a problem in their setting, of the cause and mode of transmission of LF and that hydrocele and elephantiasis had the same cause. However, this knowledge was not as prevalent amongst community drug distributors who often had misconceptions surrounding the topic.

Some articles (Kisoka et al., 2016a, 2016b; Babu and Nath, 2004; Amarillo et al., 2008; Hussain et al., 2014) discussed health worker and drug distributors' knowledge of LF and the MDA campaigns. In the Philippines (Amarillo et al., 2008), health workers considered LF to be a serious problem with those who are infected being unable to work, becoming physically deformed and stigmatised and not being able to find a partner or get married. In Tanzania (Kisoka et al., 2016b), health workers acknowledged the presence of LF due to their involvement in the MDA campaigns but did not raise it as an everyday health issue. CDDs in Tanzania felt that elephantiasis was not a problem in their communities but hydrocele was. They had conflicting views as to the cause and transmission of the disease due to differences in the training they had attended. These misconceptions were often passed onto the community during MDA campaigns when CDDs were tasked with providing information to community members (Kisoka et al., 2016a).

In India (Babu and Nath, 2004), health workers believed that mosquito control was the most important factor for controlling LF. Very few health workers knew the benefits of MDA or that the disease could be prevented using drugs (Babu and Nath, 2004). Some considered elephantiasis synonymous with LF. They mentioned that since they were not observing many lymphedema cases in their areas that they were not at risk for getting the disease (Hussain et al., 2014).

5.7 Recruitment and Training of Drug Distributors

Key message 13: The recruitment and selection of drug distributors varied across contexts. However, the majority of participants believed that the selection process should be transparent and the selection criteria clearly described in order to avoid conflict.

A few studies (Kisoka et al., 2016a; Njomo et al., 2012, 2014) discussed the recruitment and selection of drug distributors. In Kenya, CDDs were selected during Chief's meetings where the communities were also informed about the upcoming MDA campaign (Njomo et al., 2014). However, a number of CDDs believed that the process of selection was not transparent (Njomo et al., 2012). CDDs in this area were selected based on education, having volunteered in past community programmes, good behaviour and familiarity with village members. Some felt obligated to accept as many people disliked volunteer work and would not participate (Njomo et al., 2012).

In Tanzania (Kisoka et al., 2016a), the selection of CDDs was not as clear. In some settings, the community chose them. However, in other settings, influential people such as community leaders or health workers, appointed them contrary to the aim of the MDA campaigns. Selection criteria ranged from previous experience to a willingness to participate. The selection of CDDs often caused conflict in community settings. One strategy of selecting CDDs which aimed to use the existing trust in a community to foster trust in the disease elimination programme was discussed (Kisoka et al., 2016a). In smaller communities, this seemed to work effectively with people commenting that they took the drugs as someone they knew distributed them. However, in larger or urban communities where not all households knew the distributor the same strategy did not work (Kisoka et al., 2016a).

Key message 14: Studies found that in general community drug distributors did not receive adequate training to fulfil their duties during MDA campaigns.

Many studies (Kisoka et al., 2016a; Kisoka et al., 2016b; Babu and Kar, 2004; da-Costa Vroom et al., 2015; Hussain et al., 2014; Njomo et al., 2012; Njomo et al., 2014) reported findings related to the training of CDDs. Inadequate training, or in some cases no training at all, was reported as a finding in a number of studies (Hussain et al., 2014; Kisoka et al., 2016a, 2016b; Njomo et al., 2012, 2014). The following problems were highlighted: Short training session occurring too close to the distribution day did not allow for the distribution of information and messaging to the community before the start of the campaign (Hussain et al., 2014); Reported training times of 1 to 3 h over three days were inadequate to prepare distributors with the knowledge of the disease and distribution needed to complete their task (Kisoka et al., 2016a); Inadequate training lead to misinformation being passed on to the community about the MDA campaign, the cause of LF, how it is transmitted and the benefits of taking the drugs (Kisoka et al., 2016a, 2016b); When training of CDDs was perceived as inadequate by community members they were more likely to refuse to take the tablets (Kisoka et al., 2016b); many CDDs identified their training as inadequate and believed better training would have benefits to the MDA campaign (Njomo et al., 2014).

Some studies described training CDDs received including training in communication skills (Babu and Kar, 2004), the technicalities of recording distribution and calculating dosage based on measurements (da-Costa Vroom et al., 2015), the disease and its prevention (Babu and Kar, 2004), the rationale for treatment (Kisoka et al., 2016a) and the need to observe the drug being swallowed (Njomo et al., 2014). One article stated that the content of the training was consistent with WHO's recommendations (Njomo et al., 2012). When training was conducted as planned and covered communication skills, the disease and its prevention, compliance rates were higher (Babu and Kar, 2004).

5.8 Health Worker and CDD Attitudes Towards Programme Acceptability

Key message 15: Health workers and drug distributors often had positive attitudes towards the MDA campaigns viewing them as useful and reducing the problems associated with LF in their communities. Health workers and drug distributors had more negative attitudes towards the drug distribution process.

Some articles (Kisoka et al., 2016a, 2016b; Babu and Nath, 2004; Njomo et al., 2012, 2014) presented findings about the positive attitudes of health workers and drug distributors towards the MDA campaigns. In India (Babu and Nath, 2004), medical officers and NGO's viewed the programme as useful for the local people and had a generally positive attitude towards it. In Tanzania (Kisoka et al., 2016b), drug distributors felt that the programme had reduced the problems associated with LF in their communities. They believed MDA campaigns were beneficial to both individuals and the community. Local health workers felt the campaign was important and believed strongly in the drugs. They chose to walk along with distributors to ensure that they reached every house feeling it was their duty to exert some pressure on people who refused to take them (Kisoka et al., 2016a). They perceived the community as becoming more positive and accepting of the MDA campaigns. Some noted that communities were taking the initiative to request drugs from health centres. It was also noted that distributors were receiving a more open and welcome reception when arriving into communities (Kisoka et al., 2016a).

A few articles (Njomo et al., 2012; Kisoka et al., 2016a, 2016b; Babu and Nath, 2004) also presented findings of negative attitudes of health workers and drug distributors towards the MDA campaigns. Most of the negative perceptions were linked to the distribution process. Medical officers in India (Babu and Nath, 2004) were not happy with the method of distribution, especially that the drugs were distributed by non-medical personal. In Tanzania, distributors were negatively received in the community and sometimes insulted or belittled (Kisoka et al., 2016a).

Health workers and drug distributors in Tanzania and Kenya had negative perceptions of the drug distribution (Kisoka et al., 2016b; Njomo et al., 2012) including negative reactions from community members, little or no trust in their abilities from community members, difficulties in reaching communities given the terrain and the short time allocated for distribution (Kisoka et al., 2016b).

5.9 Health Worker and CDD Motivation to Participate in the MDA Campaigns

Key message 16: Health workers and CDDs were motivated to participate in MDA campaigns when they saw a decrease in the number of LF cases in their communities, an increase in community awareness and knowledge of LF, were satisfied with their jobs and felt pride in their work. Difficult physical conditions, a lack of incentives, insufficient staffing of distributors and insufficient planning and implementation of awareness activities around campaigns were demotivating.

Some articles (Amarillo et al., 2008; Baker et al., 2007; Kisoka et al., 2016b; Njomo et al., 2012) discussed factors that motivated health workers and CDDs to participate in MDA campaigns. In the Philippines (Amarillo et al., 2008), health workers were motivated to continue working with MDA due to the decrease in the number of positive cases of LF in their communities, the increase in people's awareness and knowledge about the disease and the clarification of various misconceptions regarding antifilarial drugs.

In the Dominican Republic (Baker et al., 2007), staff were motivated by increased job satisfaction after being integrated into the larger public health programme. This integration allowed them to participate in other programmes such as vaccination and due to extra support complete the MDA tasks more quickly.

In Tanzania, some distributors were motivated to participate in the MDA campaigns as they felt it improved their communication skills. They felt pride and growing self-confidence in their ability to communicate with different stakeholders. This improved communication gave them increased respect from community members as they were more clearly able to explain the rationale and benefits of the campaign (Kisoka et al., 2016a).

In Kenya, CDDs were motivated to participate when they were recognized and received items such as t-shirts as tokens of appreciation (Njomo et al., 2012).

Three articles (da-Costa Vroom et al., 2015; Kisoka et al., 2016a; Njomo et al., 2014) discussed factors that decreased motivation to participate in MDA campaigns. In Ghana, difficult physical access to communities during MDA campaigns and the lack of financial incentives decreased motivation to participate (da-Costa Vroom et al., 2015). In Tanzania, lack of financial incentives were a factor along with the limited amount of time given to distribute the drugs and the negative reception in some communities (Kisoka et al., 2016a). Finally, in Kenya, distributors found it difficult when the awareness programmes conducted before campaigns had not adequately informed the communities about the MDA campaign. They also felt that more distributors were needed to complete the MDA campaign and they wanted adequate training for their position (Njomo et al., 2012, 2014).

Key message 17: Studies found that incentives were important for drug distributor motivation when they were perceived as adequate and a demonstration of the appreciation of their time and effort. However, if the remuneration, financial or other, was perceived as not enough or less than what was offered in other public health programmes, it influenced distributor motivation and could lead to leaving the programme.

Some studies (Njomo et al., 2012; Njomo et al., 2014; Kisoka et al., 2016a; da-Costa Vroom et al., 2015; Baker et al., 2007) mentioned incentives for drug distributors. These can be classified into monetary incentives, support and items (such as T-shirts and badges). Incentives were perceived as positive when participants felt that the remuneration they received was adequate and demonstrated an appreciation of their time and effort. However, when remuneration was perceived as being too little it influenced the motivation of the distributors and could lead them to drop out of the programme. Incentives could also influence who was selected to become a CDD. In Tanzania (Kisoka et al., 2016a), some health workers feared that community leaders were appointing their relatives because of the incentives. In this case, the health workers feared that it would negatively affect the distribution and appointed community health workers instead.

Key message 18: Studies found that the majority of health workers and community drugs distributors believed that the MDA campaigns were too short and did not provide enough time to reach every household. In some cases, it was reported that there was an insufficient number of people involved in distributing tablets during MDA campaigns.

All studies found that the duration of the campaigns in their settings was too short. Studies from Ghana (da-Costa Vroom et al., 2015), India (Hussain et al., 2014) and Kenya (Njomo et al., 2012, 2014) all found that drug distributors believed the number of days allocated to distribute the drugs was too little, leading to lower rates of uptake. In Ghana (da-Costa Vroom et al., 2015) and Kenya (Njomo et al., 2012, 2014) distributors believed that longer campaigns would allow them to reach more households and adequate time to finish reporting duties.

Campaign duration ranged from one to seven days across the study sites. In Tanzania (Kisoka et al., 2016a), the community directed approach to MDA prescribes that each community should decide on the timing and duration for distribution that fits their context. However, the NTD control programme decided on a duration of three days for all contexts. This time crunch was exacerbated when campaigns happened during the rainy season when farmers were out in the fields and physical access to many areas becomes difficult.

Only one study in India (Babu and Nath, 2004) found that some doctors wanted to limit the MDA campaigns to one day in order to minimize the lower rates of compliance on subsequent days due to adverse reactions.

In Kenya (Njomo et al., 2012, 2014), an insufficient number of drug distributors was reported. In India (Hussain et al., 2014) and Tanzania (Kisoka et al., 2016b) a lack of health workers in general was reported. Participants were frustrated over the poor health care offered in their areas.

Key message 19: Studies found that there was a lack of supervision of community drug distributors during MDA campaigns.

Some studies (Babu, 2010; da-Costa Vroom et al., 2015; Njomo et al., 2012) mentioned the supervision of drug distributors. The main finding was a lack of supervision of drug distributors during MDA campaigns (Njomo et al., 2012; da-Costa Vroom et al., 2015) due to the already high workload of the supervisors (da-Costa Vroom et al., 2015). Supervisors were sometimes left to finish the reporting and tallying that was incomplete, adding to their already large workload (da-Costa Vroom et al., 2015). Health workers were most often employed to supervise the drug distribution and support CDDs with varying levels of engagement (Amarillo et al., 2008; Kisoka et al., 2016a; Njomo et al., 2012, 2014).

6. DISCUSSION

6.1 Summary of Main Results

In summary, studies found that communities lack knowledge and information about LF and the MDA campaigns and that this lack can have an impact on how many community members choose to participate in taking medication.

Studies found that often health workers had a good understanding that LF was a problem in their setting, of the cause and mode of transmission of LF and that hydrocele and elephantiasis had the same cause. However, this knowledge was not as prevalent amongst community drug distributors who often had misconceptions surrounding the topic.

Studies found that the length, timing, level of community and health system involvement, access to care for side effects, inadequate numbers of drug distributors and supervisors and motivation of drug distributors influenced participation in MDA campaigns. They also found that the inadequate training of drug distributors could influence community trust in the MDA campaign and the drug distributor themselves if there was a perception that the person was uninformed or not trained to carry out their tasks.

6.2 Agreements and Disagreements with Other Studies or Reviews

To our knowledge, there have been no other attempts to synthesize qualitative data on the feasibility and acceptance of LF MDA campaigns. Another review focused on MDA for LF drug compliance and found that some of the factors we reported here influenced drug compliance, as well (Krentel et al., 2013). They found that it was important to have well trained and motivated drug distributors as they are the ones communicating directly with the community and their actions can directly impact whether someone chooses to take the medication or not. They suggest that information is more likely to be accepted if it comes from a trusted source and that the decision to participate is affected by perceptions of the potential benefits of participation versus the risk of adverse events. In conclusion, they identified five key ingredients to increase participation. These are engender trust, tailor programmes to local conditions, take actions to minimize the impact of adverse events, promote the broader benefits of the MDA campaigns, and directly address the issue of systematic non-compliance, which harms communities by prolonging their exposure to LF (Krentel et al., 2013). The majority of these ingredients were also identified in our work and are present in our implications for practice.

Two quantitative systematic reviews reported issues related to fears of side effects as the main barrier to taking medication, as well as lack of trust or misconceptions of the MDA campaign (Krentel et al., 2013; Babu and Babu, 2014). We also found that fear of side effects and rumours circulating in the community negatively impacted the decision to participate in MDA campaigns.

Other public health programmes, such as the prevention of schistosomiasis (WHO, 2001), MDA against onchocerciasis (WHO and Edwards, 2018) and trachoma (Paho and Cogan, 2018) as well as malaria programmes to be consistent with above usage (WHO, 2017c), rely on mass administration of drugs or similar interventions (such as mass distribution of bed nets in the case of malaria) to achieve their objectives. While financial and logistic challenges are well acknowledged by stakeholders, feasibility and accessibility issues have been given much less attention.

Experiences from other preventive programmes, can also be relevant to LF. For example, a qualitative evidence synthesis regarding vaccination communication (Ames et al., 2017b) found that parents were unhappy with the timing of the information they were receiving, that they felt they did not know enough about vaccination or its side effects and this could influence their decision to vaccinate, health workers were a trusted source of information and that they wanted information tailored to their needs and context.

Parallels can also be drawn between studies of vaccination campaigns and MDA for malaria to MDA campaigns for LF as a similar approach is taken. Studies of vaccination campaigns also show a preference for community involvement in the planning and implementation of the campaigns (Oku et al., 2017, Ames et al., 2017a, Obregón and Waisbord, 2010). Similar to our findings, when communities were involved people were more willing to participate. Sterilisation rumours also played a role in people's decision not to participate in vaccination campaigns similar to those found in our rapid review (Newby et al., 2015; Ames et al., 2017a, 2017b; Feldman-Savelsberg et al., 2000).

6.3 Limitations

The methods used in the included studies may in some cases limit the applicability and completeness of the data reported. All of the included studies made use of individual or group interviews and focus group discussions as their main method of data collection. None used long-term ethnographic methods. In addition, a number of the studies were mixed methods with very limited qualitative components. While interviews and focus groups allow researchers to collect data on what people say, observational methods allow researchers to collect data about people's actions. This would have been appropriate for understanding how information is communicated before, during and after MDA campaigns and to witness what happens during the campaigns themselves. Interviews and focus group discussions seem to be the most commonly used research methods amongst qualitative researchers exploring issues related to health. This could be because they are less time-consuming than longer-term ethnographic methods. We assessed most of the included studies to have poorly reported one or more methodological domains, related to context, sampling and/or data collection methods. In addition, very few discussed researcher reflexivity.

The limitations of this rapid review include that screening, inclusion, data extraction and analysis were completed by a single author without a blind control. We did not perform a GRADE-CERQual (Lewin et al., 2015) assessment of our findings due time and funding constraints, so we cannot say the extent to which the synthesis key messages are a reasonable representation of the phenomenon of interest. However we feel that the synthesis highlights known and potentially under or unreported problems regarding acceptability and feasibility of MDA campaigns for programme managers and funding agencies and will be useful in designing and implementing future MDA campaigns. Finally, it is important to note that most of the studies included in this review were conducted in areas where not all participants accept MDA and challenges during campaigns have occurred.

6.4 Author's Conclusions

6.4.1 Implications for Practice

The following questions presented in table 3, derived from the key messages, may help programme managers and other stakeholders to assess whether the MDA campaigns they are planning adequately address the issues that are important to communities and drug distributors discussed in this rapid review.

6.4.2 Implications for Research

We believe that further high quality, qualitative research is needed on this topic. We suggest two areas for future focus. First, based on our CASP assessment of included studies there is a need for better reporting of context,

Table 3 Implications for Practice

Implications for Practice Related to Planning and Implementation	
Has an attempt been made to ...	• Ensure that adequate human resources and financial support are being given to the MDA campaigns in a timely manner? • Ensure that all levels of the health services are involved in and supportive of the planning and implementation of MDA campaigns? • Involve local community groups in planning, announcing and distribution during MDA campaigns? • Ensure that local events and context have been taken into consideration when planning the dates for and duration of MDA campaigns? The location where drugs will be distributed? • Ensure that community members have support and access to health services to deal with side effects of the medications?
Implications for Practice Related to Communication	
Has an attempt been made to ...	• Ensure that information about MDA campaigns (date, time, location, duration, medication and side effects) is communicated and tailored to each community or area in good time and often before the campaign begins? • Ensure that information about MDA campaigns is being provided to local communities in multiple formats using clear and simple text and photos in a local language? • Ensure that information about the MDA campaign is available at a wide range of health service and community settings? Is it possible to have discussions about LF and MDA in these settings? • Ensure that communication programmes are providing clear information on the cause and mode of transmission of LF? That hydrocele and elephantiasis are both caused by the same parasite? • Adjust MDA communication strategies to respond to media stories, rumours and negative publicity about MDA in order to respond to community questions and concerns that these stories may have raised?
Implications for Practice Related to Those Distributing the Drugs	
Has an attempt been made to ...	• Ensure that adequate training is being given to drug distributors on how to communicate effectively with stakeholders, the rationale and benefits of the MDA

(Continued)

Table 3 Implications for Practice—cont'd

	campaign, cause and mode of transmission of LF and side effects of the medications? • Ensure that health workers and drug distributors are able to provide clear and correct answers to questions in the communities where they work in an open and respectful way? • Ensure that drug distributors are known and trusted people in the community? • Offer incentives to drug distributors? If yes, how do they compare to the incentives given by other public health programmes?

sampling and researcher reflexivity in qualitative studies. Future studies should provide a clear description of their study context (population, setting and context related to LF) and include reflection on the researcher's role in the study and how this may have influenced the process and findings of the study. Secondly, more qualitative research needs to be done with different stakeholder groups in resistant areas or areas with lower coverage, to learn about local perceptions and community preferences in order to eliminate LF through successful MDA campaigns.

SUPPLEMENTARY DATA

Supplementary data related to this article can be found online at https://doi.org/10.1016/bs.apar.2018.09.003.

ACKNOWLEDGEMENT

We would like to acknowledge the contribution of information specialist John Eyers in creating and conducting the search for this rapid review.

REFERENCES

Amarillo, M.L.E., Belizario, V.Y., Sadiang-Abay, J.T., Sison, S.A.M., Dayag, A.M.S., 2008. Factors associated with the acceptance of mass drug administration for the elimination of lymphatic filariasis in Agusan del Sur, Philippines. Parasites Vectors 1.

Ames, H., Njang, D.M., Glenton, C., Fretheim, A., Kaufman, J., Hill, S., Oku, A., Cliff, J., Cartier, Y., Bosch-Capblanch, X., 2017a. Stakeholder perceptions of communication about vaccination in two regions of Cameroon: a qualitative case study. PLoS One 12 e0183721.

Ames, H.M., Glenton, C., Lewin, S., 2017b. Parents' and Informal Caregivers' Views and Experiences of Communication about Routine Childhood Vaccination: A Synthesis of Qualitative Evidence. The Cochrane Library.

Babu, B.V., 2010. A qualitative study on the adverse reactions of mass treatment for lymphatic filariasis in Orissa, India. Asian Pacific J. Trop. Med. 3, 55–58.

Babu, B.V., Babu, G.R., 2014. Coverage of, and compliance with, mass drug administration under the programme to eliminate lymphatic filariasis in India: a systematic review. Trans. R. Soc. Trop. Med. Hyg. 108, 538–549.

Babu, B.V., Kar, S.K., 2004. Coverage, compliance and some operational issues of mass drug administration during the programme to eliminate lymphatic filariasis in Orissa, India. Trop. Med. Int. Health 9, 702–709.

Babu, B.V., Nath, N., 2004. The programme to eliminate lymphatic filariasis in Orissa, India: the attitudes of some programme partners. Ann. Trop. Med. Parasitol. 98, 751–756.

Baker, M.C., Mcfarland, D.A., Gonzales, M., Diaz, M.J., Molyneux, D.H., 2007. The impact of integrating the elimination programme for lymphatic filariasis into primary health care in the Dominican Republic. Int. J. Health Plann. Manag. 22, 337–352.

Carlsen, B., Glenton, C., Pope, C., 2007. Thou shalt versus thou shalt not: a meta-synthesis of GPs' attitudes to clinical practice guidelines. Br. J. Gen. Pract. 57, 971–978.

Critical Appraisal Skills Programme, 2017. CASP Qualitative Checklist.

Da-Costa Vroom, F.B., Aryeetey, R., Boateng, R., Anto, F., Aikins, M., Gyapong, M., Gyapong, J., 2015. Data reporting constraints for the lymphatic filariasis mass drug administration activities in two districts in Ghana: a qualitative study. SAGE Open Medicine 3, 1–9.

Feldman-Savelsberg, P., Ndonko, F.T., Schmidt-Ehry, B., 2000. Sterilizing vaccines or the politics of the womb: retrospective study of a rumor in Cameroon. Med. Anthropol. Q. 14, 159–179.

Glenton, C., Colvin, C.J., Carlsen, B., Swartz, A., Lewin, S., Noyes, J., Rashidian, A., 2013a. Barriers and facilitators to the implementation of lay health worker programmes to improve access to maternal and child health: qualitative evidence synthesis. Cochrane Lib. 10.

Glenton, C., Colvin, C.J., Carlsen, B., Swartz, A., Lewin, S., Noyes, J., Rashidian, A., 2013b. Barriers and facilitators to the implementation of lay health worker programmes to improve access to maternal and child health: qualitative evidence synthesis. Cochrane Database Syst. Rev. 10.

Glenton, C., Lewin, S., Gülmezoglu, A.M., 2016. Expanding the evidence base for global recommendations on health systems: strengths and challenges of the OptimizeMNH guidance process. Implement. Sci. 11, 98.

Gopinathan, U., Lewin, S., Glenton, C., 2014. Implementing large-scale programmes to optimise the health workforce in low-and middle-income settings: a multicountry case study synthesis. Trop. Med. Int. Health 19, 1437–1456.

Gyapong, J.O., Owusu, I.O., Vroom, F.B.D.-C., Mensah, E.O., Gyapong, M., 2018. Elimination of lymphatic filariasis: current perspectives on mass drug administration. Res. Rep. Trop. Med. 9, 25.

Haby, M.M., Chapman, E., Clark, R., Barreto, J., Reveiz, L., Lavis, J.N., 2016. What are the best methodologies for rapid reviews of the research evidence for evidence-informed decision making in health policy and practice: a rapid review. Health Res. Policy Sys. 14, 83.

Hussain, M.A., Sitha, A.K., Swain, S., Kadam, S., Pati, S., 2014. Mass drug administration for lymphatic filariasis elimination in a coastal state of India: a study on barriers to coverage and compliance. Infect Dis Poverty 3.

Ichimori, K., King, J.D., Engels, D., Yajima, A., Mikhailov, A., Lammie, P., Ottesen, E.A., 2014. Global programme to eliminate lymphatic filariasis: the processes underlying programme success. PLoS Negl. Trop. Dis. 8, e3328.

Irvine, M.A., Stolk, W.A., Smith, M.E., Subramanian, S., Singh, B.K., Weil, G.J., Michael, E., Hollingsworth, T.D., 2017. Effectiveness of a triple-drug regimen for global elimination of lymphatic filariasis: a modelling study. Lancet Infect. Dis. 17, 451–458.

Kastner, R.J., Stone, C.M., Steinmann, P., Tanner, M., Tediosi, F., 2015. What is needed to eradicate lymphatic filariasis? A model-based assessment on the impact of scaling up mass drug administration programs. PLoS Negl. Trop. Dis. 9 e0004147.

King, J.D., Zielinski-Gutierrez, E., PA'au, M., Lammie, P., 2011. Improving community participation to eliminate lymphatic filariasis in American Samoa. Acta Trop. 120, S48–S54.

Kisoka, W., Mushi, D., Meyrowitsch, D.W., Malecela, M., Simonsen, P.E., Tersbøl, B.P., 2016a. Dilemmas of community-directed mass drug administration for lymphatic filariasis control: a qualitative study from urban and rural Tanzania. J. Biosoc. Sci. 1–16.

Kisoka, W.J., Tersbøl, B.P., Meyrowitsch, D.W., Simonsen, P.E., Mushi, D.L., 2016b. Community members' perceptions of mass drug administration for control of lymphatic filariasis in rural and urban Tanzania. J. Biosoc. Sci. 48, 94–112.

Krentel, A., Aunger, R., 2012 Aug. Causal chain mapping: a novel method to analyse treatment compliance decisions relating to lymphatic filariasis elimination in Alor, Indonesia. Health Policy Plan. 27 (5), 384–395. https://doi.org/10.1093/heapol/czr048. Epub 2011 Jun 28.

Krentel, A., Fischer, P.U., Weil, G.J., 2013. A review of factors that influence individual compliance with mass drug administration for elimination of lymphatic filariasis. PLoS Negl. Trop. Dis. 7, e2447.

Lewin, S., Munabi-Babigumira, S., Glenton, C., Daniels, K., Bosch-Capblanch, X., Van Wyk, B.E., Odgaard-Jensen, J., Johansen, M., Aja, G.N., Zwarenstein, M., 2010 Mar. Lay health workers in primary and community health care for maternal and child health and the management of infectious diseases. Cochrane Database Syst. Rev. 17 (3), CD004015. https://doi.org/10.1002/14651858.CD004015.pub3.

Lewin, S., Glenton, C., Munthe-Kaas, H., Carlsen, B., Colvin, C.J., Gülmezoglu, M., Noyes, J., Booth, A., Garside, R., Rashidian, A., 2015. Using qualitative evidence in decision making for health and social interventions: an approach to assess confidence in findings from qualitative evidence syntheses (GRADE-CERQual). PLoS Med. 12 e1001895.

Michael, E., Bundy, D., Grenfell, B., 1996. Re-assessing the global prevalence and distribution of lymphatic filariasis. Parasitology 112, 409–428.

Muloliwa, A., Cartier, Y., Ames, H., Oku, A., Glenton, C., Cliff, J., Hill, S., Oyo-Ita, A., Rada, G., Bosch-Capblanch, X., Kaufman, J., S, L., 2017. Health System Barriers and Facilitators to Scaling up Communication for Childhood Vaccination in Low-income Settings: A Country Case Synthesis (Unpublished).

Munro, S.A., Lewin, S.A., Smith, H.J., Engel, M.E., Fretheim, A., Volmink, J., 2007. Patient adherence to tuberculosis treatment: a systematic review of qualitative research. PLoS Medicine 4, e238.

Newby, G., Hwang, J., Koita, K., Chen, I., Greenwood, B., VON Seidlein, L., Shanks, G.D., Slutsker, L., Kachur, S.P., Wegbreit, J., 2015. Review of mass drug administration for malaria and its operational challenges. Am. J. Trop. Med. Hyg. 93, 125–134.

Njomo, D.W., Amuyunzu-Nyamongo, M., Magambo, J.K., Ngure, P.K., Njenga, S.M., 2012. Factors associated with the motivation of community drug distributors in the Lymphatic Filariasis Elimination Programme in Kenya. South African J. Epidemiol. Infect. 27, 66–70.

Njomo, D.W., Mukoko, D.A., Nyamongo, N.K., Karanja, J., 2014. Increasing coverage in mass drug administration for lymphatic filariasis elimination in an urban setting: a study of Malindi Town, Kenya. PLoS One 9.

Noyes, J., Popay, J., Pearson, A., Hannes, K., 2008. 20 Qualitative Research and Cochrane Reviews. In: Cochrane Handbook for Systematic Reviews of Interventions.

Obregón, R., Waisbord, S., 2010. The complexity of social mobilization in health communication: top-down and bottom-up experiences in polio eradication. J. Health Commun. 15, 25–47.

Oku, A., Oyo-ITA, A., Glenton, C., Fretheim, A., Eteng, G., Ames, H., Muloliwa, A., Kaufman, J., Hill, S., Cliff, J., Cartier, Y., Bosch-Capblanch, X., Rada, G., S, L., 2017. Factors affecting the implementation of childhood vaccination communication strategies in Nigeria: a qualitative study. BMC Publ. Health 17.

Paho, Cogan, J.E., 2018. Trachoma - Key facts. Available: http://www.who.int/news-room/fact-sheets/detail/trachoma.

Reuters, T., 2016. EndNote™. X7.7.1 (Bld 10036) Ed.

Supporting the Use of Research Evidence Collaboration, 2011. SURE Guides for Preparing and Using Evidence-based Policy Briefs: 5 Identifying and Addressing Barriers to Implementing Policy Options. Version.

The World Bank Data; Coun, 2016 The World Bank Data; Countries and economies [Online]. Available: http://data.worldbank.org/country.

Wolrd Health Organization, 2011. Monitoring and Epidemiological Assessment of Mass Drug Administration in the Global Programme to Eliminate Lymphatic Filariasis: A Manual for National Elimination Programmes.

Wolrd Health Organization, 2017b. Guideline: Alternative Mass Drug Administration Regimens to Eliminate Lymphatic Filariasis. World Health Organization, Geneva.

World Health Organization & Edwards, M., 2018. Onchocerciasis - Key Facts. Available: http://www.who.int/news-room/fact-sheets/detail/onchocerciasis.

World Health Organization, 2001. Expert Committee on the Control of Schistosomiasis: Prevention and Control of Schistosomiasis and Soil-transmitted Helminthiasis. In: WHO Technical Report Series. World Health Organization, Geneva.

World Health Organization, 2016. Global Health Estimates 2015: Disease Burden by Cause, Age, Sex, by Country and by Region, 2000–2015. Geneva.

World Health Organization, 2017a. Global program to eliminate lymphatic filariasis: progress report 2016. Wkly. Epidemiol. Rec. 92, 592–607.

World Health Organization, 2014. 15. Using evidence from qualitative research to develop WHO guidelines. In: WHO (Ed.), WHO Handbook for Guideline Development, second ed. WHO, Geneva.

World Helath Organization, 2017c. World Malaria Report 2017. World Health Organization, Geneva.

Wynd, S., Carron, J., Selve, B., Leggat, P.A., Melrose, W., Durrheim, D.N., 2007. Qualitative analysis of the impact of a lymphatic filariasis elimination programme using mass drug administration on Misima Island, Papua New Guinea. Filaria J. 6.

CHAPTER SEVEN

Mobile Phone Devices and Handheld Microscopes as Diagnostic Platforms for Malaria and Neglected Tropical Diseases (NTDs) in Low-Resource Settings: A Systematic Review, Historical Perspective and Future Outlook

Alon Vasiman*, J. Russell Stothard§ and Isaac I. Bogoch*,¶,1
*Department of Medicine, University of Toronto, Toronto, ON, Canada
§Department of Parasitology, Liverpool School of Tropical Medicine, Liverpool, United Kingdom
¶Divisions of General Internal Medicine and Infectious Diseases, Toronto General Hospital, University Health Network, Toronto, ON, Canada
[1]Corresponding author: E-mail: isaac.bogoch@uhn.ca

Contents

Advances in Parasitology, Volume 103
ISSN 0065-308X
https://doi.org/10.1016/bs.apar.2018.09.001

Abstract

The accurate, rapid, and cost-effective diagnosis of malaria and neglected tropical diseases (NTDs) in low-resource settings may benefit by significant technological advances in handheld and mobile phone microscopy. We systematically review the available literature in this field and discuss the future directions in which these technologies may be applied. English-language studies from the PubMed, Embase, and Web of Sciences were searched through April 2018 for observational and interventional studies reporting diagnostic characteristics of handheld and mobile phone microscopy devices as compared to field-established gold standard reference tests. Seventeen studies were included in the analysis. Findings included the high performance of the Newton Nm1 microscope in the diagnosis of *Plasmodium* species, *Schistosoma mansoni*, and soil-transmitted helminths (STHs), exhibiting sensitivity and specificity values often greater than 90%. Similarly, the CellScope was shown to have excellent diagnostic characteristics in the detection of *Loa loa* and *Schistosoma* species. Fluorescent microscopy was found to have high specificity and sensitivity in the diagnosis of *Plasmodium* species.

Mobile phone technologies and handheld microscopes hold significant promise in the rapid and effective diagnosis of malaria and NTDs in areas where accurate diagnosis is vital. Although many of these technologies have yet to be securely embedded within the health system and studied directly in this context, the foundations for significant healthcare advances and impact have already been laid by several studies conducted within the last decade.

1. INTRODUCTION

Malaria and neglected tropical diseases (NTDs) continue to cause significant morbidity and mortality worldwide (Murray et al., 2012; Snow et al., 2017; Vos et al., 2017). Several interventions are needed to combat these diseases effectively including integrating effective and low-cost diagnostic tools into routine clinical and public health practice. However, routine diagnosis of malaria and NTDs remains challenging due to the paucity of specialized equipment and need for specialist or highly trained individuals in many low-resource settings. To tackle these complex issues, implementation research involving several complimentary approaches is needed, including publicly and privately funded ventures to develop new inexpensive diagnostic platforms, tools, and protocols that improve diagnostic accuracy for common medical conditions of clinical and public health significance. Within this realm, there have been several recent developments involving handheld microscopes and mobile phone microscopes (Coulibaly et al., 2016a; Rajchgot et al., 2017). Some advantages of these technologies in low-resource settings include low handling costs, portability, battery power, image capture with telemedicine possibilities, and low equipment training requirements.

Specific to portable microscopy, mobile phone devices converted into microscopes provide a distinct advantage due their built-in high-resolution cameras which are continually being improved and refined. Furthermore, mobile phones are already widely used even in low-resource settings and their use is familiar even to un-trained personnel (Bastawrous and Armstrong, 2013). Mobile phones can also be modified or customized to act as microscopes with newly developed applications or added-on hardware. These features, along with increased storage space, access to powerful computing capabilities, connectivity with global positioning system (GPS), and improved access to the internet provide mobile phones with numerous appealing features that make them ideal platforms in low-resource settings.

While there were some early innovations in handheld microscopes (Meade Instruments, 2018), recent innovations and improvement in the realms of optics, design, and engineering have produced devices with the potential to be effective in real-world clinical and public health practice in low-resource settings. The field remains in its early stages of development and continues to see a rapid evolution of novel devices harnessing creative methods to image pathogens of global health significance. In this article, we first outline a brief history of handheld microscopy. We then present the first systematic review of handheld and mobile phone microscope technologies to evaluate their effectiveness, compared to gold-standard diagnostic methods for NTD diagnosis in low-resource clinical settings. We also summarise future applications of mobile technologies that have been tested in laboratory settings but have not yet been examined in the field.

2. HISTORICAL PERSPECTIVE

The growth of mobile and handheld devices in the field of microscopy has mirrored the rapid development of diagnostics and treatment in the field of parasitology (Stothard and Rollinson, 2018). The history of portable microscopy with traditional light optics can be traced back over seventy-five years to a particularly influential clinical pioneer named John Norris McArthur (b. 1901 d. 1996) who promoted their application in parasitic disease detection and diagnosis. While a medical student in London shortly before WWII, McArthur conceived and designed a particularly revolutionary format with inverted optics and carefully placed prisms that radically shortened the reflected light path. First fashioned inside wood and then later metal housings, these devices were essentially self-contained monocular microscopes small enough (10.2 cm × 6.4 cm × 5.1 cm) to be held in one hand and carried inside a coat pocket (Dunning and Stothard, 2007).

The McArthur microscope and its derivations (Fig. 1), which bore his name, could deliver as much optical performance with more versatility than the traditional bench-top compound microscope. Originally fashioned to use external natural light and later an internal bulb from either a battery or mains supply, the McArthur microscope set the industry standard for portable microscopy for several decades. Its use by UK-based medical practitioners such as Murray Longmore advanced the concept of 'bedside' microscopy alongside miniaturization of diagnostic testing and streamlined staining protocols (Longmore, 1983, 1986). A particularly seminal study by Collier and Longmore (1983) assessed the performance of the McArthur microscope in the field-diagnosis of malaria in the Solomon Islands (Collier and Longmore, 1983). Whilst proven useful, MacArthur's microscope fell out of regular commercial production upon his death largely due the loss of its strongest advocate and its increasing expense owing to individual-tooling.

The late 1980s saw advances in computer design and mouldable plastic technology leading to the original ambition of the McArthur microscope being reinvigorated. This included the production and commercial retail of low-cost portable microscopes such as the 'Enhelion' in 1988 originally

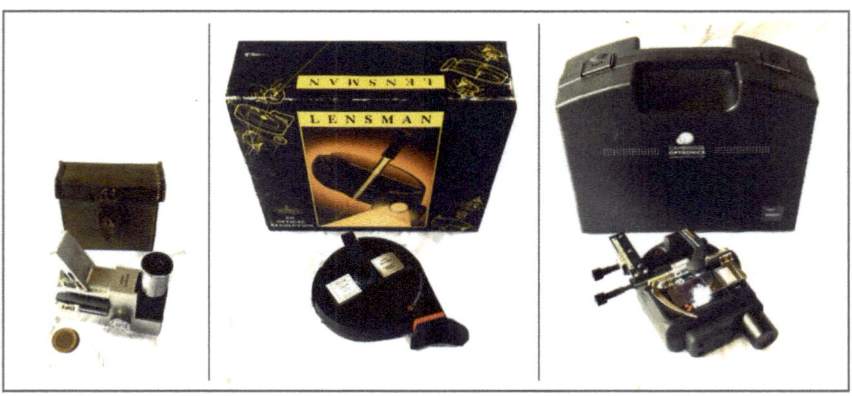

Figure 1 Three handheld microscopes illustrative of advances in microscope design and tooling beginning with the McArthur (left), Lensman (centre) and Newton Nm1 (right), with the carry box of each depicted. For scale, the coin seen on the left image is a €1 coin.

priced at £99 UK with magnifications of ×80 and ×200. This unit was produced by Science of Cambridge, conceived and designed by Keith Dunning and Rick Dickinson (Kreindler, 2013a,b). Over the next two decades, there were several fluctuations in supply, demand, and commercial production of similar portable microscopes (Kreindler, 2013a). However, the field of portable microscopy within clinical application was significantly advanced by an initiative from the Wellcome Trust in 2009 to make the Newton Nm1 microscope. This device used light emitting diode (LED) technology, which offered magnifications at ×100, ×400, and even higher at ×600 (dry) and ×1000 (oil emersion), for application in peripheral health clinics in low and middle income countries (Kreindler, 2013b).

3. SYSTEMATIC REVIEW METHODS

Reporting conformed to PRISMA (Preferred Reporting Items for Systematic Reviews and Meta-Analyses) guidelines (Moher et al., 2009). The study's protocol has not been published or registered elsewhere.

4. DATA SOURCES AND SEARCHES

We searched the Web of Science, PubMed, and Embase from inception through 15 April 2018, including only English-language literature. The key concepts included in our search strategies were mobile phone and handheld devices, tropical infections and NTDs, and microscopy (see Supplemental Table 1 for details). We excluded conference proceedings. We also conducted hand search reviews of literature cited by relevant articles and reviews.

5. STUDY SELECTION

Two investigators (AV and IB) independently screened titles and abstracts. We included interventional and observational studies that compared a handheld, point-of-care, or mobile phone diagnostic tool with a reference or gold standard diagnostic tool in the diagnosis of a NTD and malaria. We also only included studies that were conducted in low-resource field settings. Demonstration studies, and/or studies conducted in high-resource settings were excluded. Two investigators then independently evaluated the full-text articles and disagreements were resolved by consensus.

6. DATA EXTRACTION AND QUALITY ASSESSMENT

One investigator extracted study data, including setting, population, the diagnostic measures, and descriptions of the mobile technology. The primary outcomes were the sensitivity and specificity, including CIs, of the diagnostic tool of interest compared to a gold standard. Positive and negative predictive values were also extracted from the studies. Due to significant study heterogeneity in terms of the infections being investigated, the types of diagnostic tools, and reference standards, we did not perform a meta-analysis. One reviewer independently assigned each study a level for risk of bias (low, high, or unclear) on the basis of the Quality Assessment of Diagnostic Accuracy Studies (QUADAS2) criteria (Whiting, 2011) (Supplemental Table 2). The four domains of bias that were judged were patient selection, index testing, reference standard, and flow and timing. Studies that randomly selected their samples from a representative population were deemed low risk of bias. Similarly, studies that chose well-defined and well-established testing methods as their gold standard comparison were deemed low risk of bias. Lastly, studies that blinded evaluators and had multiple evaluators were also deemed to be low risk from bias according to these criteria.

7. RESULTS

With removal of duplicate records, the search strategies identified 462 studies, 71 of which we evaluated as full-text articles (Fig. 2). We chose a total of 17 reports in the final selection for inclusion in the review (Table 1). All studies except for one (Bogoch et al., 2016) were conducted in Africa. Eight of the studies were conducted exclusively in a paediatric population, the most common subgroup to be studied (Stothard et al., 2005; Nkrumah et al., 2011; Bogoch et al., 2013, 2014a; b; Ephraim et al., 2015; Coulibaly et al., 2016a; Bogoch et al., 2017a). The most common organisms studied were *Plasmodium* species (Hassan et al., 2010, 2011; Sousa-Figueiredo et al., 2010; Nkrumah et al., 2011; Birhanie, 2016), and the most commonly studied mobile technologies were the Newton Nm1 portable microscope (Stothard et al., 2014; Bogoch et al., 2014b, 2016; Coulibaly et al., 2016a; b) and Cyscope (Hassan et al., 2010, 2011; Sousa-Figueiredo et al., 2010; Nkrumah et al., 2011; Birhanie, 2016). Gold standard comparisons were most commonly with conventional light microscopy using

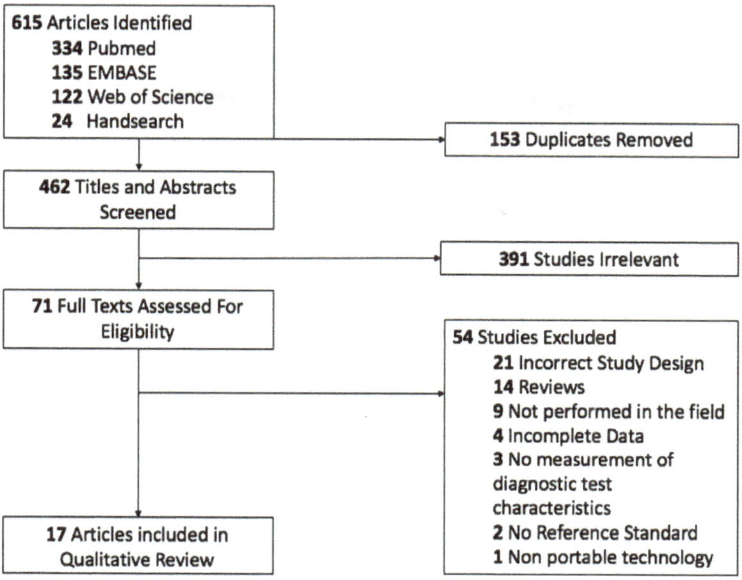

Figure 2 Flow chart of study selection for diagnostic devices used to diagnose neglected tropical diseases in low-resource settings.

standard Kato-Katz thick smear tests (Katz et al., 1972) for helminth and thick Giemsa stain smears for studies examining malaria. The number of samples studied in the selected papers ranged from 33 to 16 259. The number of reviewers/microscopists for each study ranged from 1 to 4 with an average of 1.25 reviewers per study. The technologies roughly divided into four categories: first-generation ball lens-mounted mobile phone devices, second-generation mobile phone devices, handheld light microscopes, and the Cyscope.

7.1 First Generation Ball Lens-Mounted Mobile Phone Devices

The results of two studies using first generation ball lens-mounted devices for microscopy are shown in Table 2. These studies (Bogoch et al., 2013, 2014a) developed a novel tool for portable microscopy by mounting a 3-mm ball lens onto a mobile phone. This was used to diagnose *Schistosoma haematobium* eggs in the urine that resulted in poor sensitivity for low intensity infections (29%) and moderate sensitivity for high intensity infections (78.3%). For all infection intensities, the specificity was moderate at 64%

Table 1 Overview of the Studies Included in the Qualitative Review, by Device Type and Year

	Setting	Number of Samples Tested	Population	Organism	Mobile Technology	Gold Standard	Reviewers
Bogoch et al., 2013	Pemba Island, Tanzania	199	Paediatric	*T. trichura, A. lumbricoides*	iPhone add-on	Kato Katz Method, Olympus CX21 microscope	1
Bogoch et al., 2014a	Azaguie, Côte d'Ivoire	180	Paediatric	*Schistosomiasis* spp., *T. trichura*	iPhone add-on, Newton Nm1	Kato Katz Method, Olympus CX21 microscope	1
Bogoch et al., 2017a	Central Region, Ghana	60	Paediatric	*S. haematobium*	Novel Mobile phone device	Olympus CX21 microscope	1
Bogoch et al., 2014b	Pemba Island, Tanzania	182	Paediatric	*T. trichura, A. lumbricoides*	Newton Nm1	Kato Katz Method	1
Stothard et al., 2014	Uganda	50	Women and children	*Plasmodium* spp,	Newton Nm1	Olympus CX22 microscope	4
Coulibaly et al., 2016b	Grand Moutcho, Côte d'Ivoire	223	General	*P. falciparum*	Newton Nm1	Olympus CX22 microscope	2
Bogoch et al., 2016	Laos	104	Adult	*O. viverrini*	Newton Nm1	Kato Katz Method, Olympus CX21 microscope	1

Coulibaly et al., 2016a	Grand Moutcho, Côte d'Ivoire	226	Paediatric	*S.hematobium/ mansoni, G. intestinalis, E. histolytica/ dispar,*	Newton Nm1, reversed lens CellScope	Kato Katz Method, Olympus CX21 microscope	Unclear
Stothard et al., 2005	Hoima and Mayuge districts, Uganda	685	Paediatric	*Schistosoma* spp	Meade Readview	Kato Katz Method	4
Ephraim et al., 2015	Central Region, Ghana	49	Paediatric	*Schistosoma* spp	CellScope and Foldscope	Conventional light microscopy	Unclear
D'Ambrosio et al., 2015	Cameroon	33	General	*Loa loa*	CellScope Loa	Giemsa Stained Thick Smear	2
Kamgno et al., 2017	Okola District, Cameroon	16,259	General	*Loa loa*	LoaScope	Thick Smear	2
Sousa-Figueiredo et al., 2010	Uganda	1,530	Women and children	*Plasmodium* spp.	Cyscope	Thick Giemsa Smear	Unclear
Hassan et al., 2010	Sinnar, Sudan	293	Adult	*P. falciparum*	Cyscope	Thick Giemsa Smear	1
Hassan et al., 2011	Central Region, Sudan	128	Pregnant Women	*P. falciparum*	Cyscope	Thick Giemsa Smear	1
Nkrumah et al., 2011	Ashanti Region, Ghana	263	Paediatric	*P. falciparum*	Cyscope	Thick Giemsa Smear	2
Birhanie, 2016	Metema District, Ethiopia	180	General	*Plasmodium* spp.	Cyscope	Thick Giemsa Smear	1

Table 2 Summary of Diagnostic Characteristics of the Devices Studied

	Organism	Sensitivity (95%CI)	Specificity (95%CI)	PPV (95%CI)	NPV (95%CI)	Pearson R	Device
First-Generation Ball Lens-Mounted Mobile Phone Devices							
Bogoch et al., 2014a	*S. mansoni* (all infections)	68.2 (60.1–75.5)	64.3 (35.1–87.2)	95.4 (89.5–98.5)	15.8 (7.5–27.9)	NR	iPhone Add-on
	S. mansoni (low intensity infection)	29 (14.9–48.2)	NR	NR	NR	NR	iPhone Add-on
	S. mansoni (high intensity infection)	78.3 (70–85.1)	NR	NR	NR	NR	iPhone Add-on
	T. trichiura (all infections)	30.8 (19.9–43.4)	71 (61.1–79.6)	40.8 (27–55.8)	61.2 (51.7–70.1)	NR	iPhone Add-on
	T. trichiura (low intensity infection)	26.3 (15.9–39.9)	NR	NR	NR	NR	iPhone Add-on
	T. trichiura (moderate/heavy intensity infection)	62.5 (25.9–89.9)	NR	NR	NR	NR	iPhone Add-on
Bogoch et al., 2013	STHs (all infections)	69.4 (61.8–76)	61.5 (40.7–79.1)	92.3 (85.9–96)	23.2 (14.2–35.2)	NR	iPhone Add-on
	A. lumbricoides (all infections)	81 (65.4–90.9)	87.3 (80.7–91.9)	63 (48.7–75.4)	94.5 (89.1–97.4)	NR	iPhone Add-on
	A. lumbricoides (low intensity infection)	74.5 (53.4–88.1)	NR	NR	NR	NR	iPhone Add-on
	A. lumbricoides (moderate/heavy intensityinfection)	93.3 (66–99.7)	NR	NR	NR	NR	iPhone Add-on
	T. trichiura (all infections)	54.4 (46.3–62.3)	63.4 (46.9–77.4)	85.1 (76.4–91.2)	26.5 (18.4–36.6)	NR	iPhone Add-on

	T. trichiura (low intensity infection)	43.9 (34.5–53.8)	NR	NR	NR	NR	iPhone Add-on
	T. trichiura (moderate/heavy intensityinfection)	76.5 (62.2–86.7)	NR	NR	NR	NR	iPhone Add-on
	Hookworm (all infections)	14.3 (8.3–23.1)	89.1 (81–94.2)	56 (35.3–75)	51.7 (44.1–59.3)	NR	iPhone Add-on
Second-Generation Mobile Phone Devices							
Ephraim et al., 2015	*S. haematobium*	55.9 (38.1–72.4)	93 (66–99.7)	NR	NR	NR	Foldscope
	S. haematobium	67.6 (49.4–82)	100 (74.7–100)	NR	NR	NR	CellScope
D'Ambrosio et al., 2015	*Loa loa*	100	94	NR	NR	0.96	CellScope
Coulibaly et al., 2016a	*S. mansoni*	50 (25.4–74.6)	99.5 (97–100)	85.7 (42–99.2)	97.3 (93.9–98.9)	NR	CellScope
	S. haematobium	35.6 (25.9–46.4)	100 (96.6–100)	100 (86.7–100)	70.1 (63.1–76.3)	0.92	CellScope
Kamgno et al., 2017 [1]	*Loa loa*	92.8	99.7 (99.6–99.8)	92.8	99.7 (99.6–99.8)	NR	LoaScope
Bogoch et al., 2017a	*S. haematobium* (all infection)	100 (59.8–100)	NR	NR	NR	NR	Novel Mobile Phone Device
	S. haematobium low intensity infection	72.1 (56.1–84.1)	100 (75.9–100)	100 (86.3–100)	57.1 (37.4–75)	NR	Novel Mobile Phone Device
	S. haematobium (high intensity infection)	65.7 (47.7–80.3)	NR	NR	NR	NR	Novel Mobile Phone Device

Handheld Microscopes With and Without Mobile Phone Attachments						
Bogoch et al., 2014b	*A. lumbricoides*	99.2	96.4	98.4	98.2	0.90
	T. trichiura	93.8	81	97.4	63	0.94
Stothard et al., 2014	*Plasmodium* spp.	93.5 (78.6–99.2)	100 (82.4–100)	100 (88.1–100)	90.5 (69.6–98.8)	0.98
Coulibaly et al., 2016b	*P. falciparum*	80.2 (73.1–85.9)	100 (92.6–100)	100 (96.4–100)	65.6 (54.9–74.9)	0.997
Bogoch et al., 2016	*Opisthorchis viverrini*	70.6 (59.6–79.7)	89.5 (65.5–98.2)	96.8 (87.8–99.4)	40.5 (26–56.7)	0.98
Bogoch et al., 2014a	*S. mansoni* (all infections)	84.8 (78–90.1)	85.7 (57.2–98.2)	98.5 (94.6–99.8)	34.3 (19.1–52.2)	NR
	S. mansoni (low infection intensity)	45.2 (27.8–63.7)	NR	NR	NR	NR
	S. mansoni (high infection intensity)	95 (89–98)	NR	NR	NR	NR
	T. trichiura (all infections)	81.5 (70–90.1)	93 (86.1–97.1)	88.3 (77.4–95.2)	88.6 (80.9–94)	NR
	T. trichiura (low infection intensity)	80.7 (67.7–89.5)	NR	NR	NR	NR
	T. trichiura (moderate/heavy infection intensity)	87.5 (46.7–99.3)	NR	NR	NR	NR
	S. haematobium (all infection intensity)	78.6 (49.2–95.3)	91 (85.5–94.9)	42.3 (23.4–63.1)	98.1 (94.4–99.6)	NR
	S. haematobium (low infection intensity)	72.7 (39.3–92.7)	NR	NR	NR	NR
	S. haematobium (high infection intensity)	100	NR	NR	NR	NR

Coulibaly et al., 2016a	*S. mansoni*	91.7 (59.8–99.6)	99.5 (97–100)	91.7 (59.8–99.6)	99.5 (97–100)	NR
	S. haematobium	81.1 (71.2–88.3)	97.1 (92.2–99.1)	94.8 (86.5–98.3)	88.6 (82.1–93)	0.98
	E. histolytica/dispar	83.3 (36.5–99.1)	100 (96–100)	100 (46.3–100)	99.1 (94.6–100)	NR
	G. intestinalis	84 (63.1–94.7)	100 (95.2–100)	100 (80.8–100)	96 (89.5–98.7)	NR
Stothard et al., 2005	*S. mansoni*	85	96	95	88	NR
CyScope						
Sousa-Figueiredo et al., 2010	*P. falciparum*, Adults	86.7 (79.3–92.2)	38.8 (33.6–44.1)	32.8 (27.7–38.3)	89.4 (83.4–93.8)	NR
	P. falciparum, Children	92.1 (89.6–94.1)	28.6 (22.8–34.9)	77.1 (73.9–80.2)	57.9 (48.3–67.1)	NR
Hassan et al., 2010	*P. falciparum*, Adults	98.2 (90.6–100)	98.3 (95.7–99.5)	93.3 (83.8–98.2)	99.6 (97.6–100)	NR
Hassan et al., 2011	*P. falciparum*, Pregnant women	97.6 (92.2–99.6)	89.1 (77.5–95.9)	94.1 (87.4–97.8)	95.3 (85.4–99.2)	NR
Nkrumah et al., 2011	*P. falciparum*, Children	100 (96.6–100)	97.4 (93.6–99.3)	96.4 (91–99)	100 (97.6–100)	NR
Birhanie, 2016	*Plasmodium* spp., Adults and Children	93.8 (87.1–100)	87.9 (79.7–96.1)	86.4 (77.2–95.5)	94.6 (88.7–100)	NR

CI, confidence interval; *NPV*, Negative predictive value; *NR*, Not reported; *PPV*, Positive predictive value.
[1]Threshold for positive *L. loa* microfilariae was a density below 20,000 mf per milliliter of blood.

with high positive predictive value but very poor negative predictive value. In these two studies, the test characteristics were poor for the diagnosis of *T. trichiura* – particularly the sensitivity (30%–55%) and negative predictive value (26%–61%).

7.2 Second Generation Mobile Phone Devices

This group of mobile phone-related devices included the reverse-lens CellScope and Foldscope devices as well as other 3D printed devices (Table 2). The reversed-lens CellScope is a microscope that is operated in conjunction with an unmodified mobile phone. Users of the device place the lens such that it aligns with the camera of the mobile phone, and capture images by holding the mobile phone microscope directly above a sample and manually adjusting the focus. Use of this device was studied for the diagnosis of *Schistosoma* species by two studies (Ephraim et al., 2015; Coulibaly et al., 2016a), which found that the tool provided a high degree of specificity (upwards of 99%) and had a high negative predictive value, indicating a very high proportion of true negatives. A modified version of the CellScope, now known as the LoaScope, was studied in the diagnosis of *Loa loa* in two other investigations (D'Ambrosio et al., 2015; Kamgno et al., 2017). The LoaScope uses a similar modified lens that is attached to mobile phones but also provides the benefit of automatically quantifying microfilariae in the collected samples. Notably, the LoaScope had a sensitivity ranging from 93–100% and specificity close to 100%. The Foldscope, a device that is similarly added to a mobile phone, consists of a paper device harnessing a 2.38 mm ball lens and an LED. Ephraim et al. (Ephraim et al., 2015) investigated the use of this device for the diagnosis of *Schistosoma haematobium* in a paediatric population in Ghana. Compared to the CellScope, the Foldscope had both poorer sensitivity (56% vs. 68%) and specificity (93% vs. 100%).

The third technology using a novel mobile phone design (Bogoch et al., 2017a) focused on the diagnosis of *Schistosoma haematobium* in rural Ghana. Here, the authors used a lightweight 3D printed custom-designed optomechanical unit that was attached to the camera unit on a mobile phone. The regular mobile phone camera application was used to capture images of the sample. Although the sensitivity of the device was only 72%, the specificity and positive predictive value were estimated to be 100%. Thus, there were no false positives in the study.

7.3 Handheld Light Microscopes Used With and Without Mobile Phone Attachments

The best studied of the portable microscopy devices is the Newton Nm1 microscope. This monocular light microscope is a commercially available and lightweight (~0.5 kg) device that can be secured onto mobile phones. Similar to the CellScope and Foldscope, this technology relies on the mobile phone's camera to examine specimens on slides. As seen in Table 2, a wide variety of organisms have been studied using this device. Of note, the microscope performed well with regard to the diagnosis of *Plasmodium* species (Stothard et al., 2014), *A. lumbrocoides* and *T. trichiura* (Bogoch et al., 2014b), and *Schistosoma mansoni* (Bogoch et al., 2014a; Coulibaly et al., 2016a) infections, exhibiting sensitivity and specificity values often greater than 90%. Furthermore, the device appears to have high specificity but poor sensitivity for the diagnosis of intestinal protozoa (Coulibaly et al., 2016a).

A second device, the Meade Readview handheld microscope, is an inexpensive and lightweight monocular microscope that is portable and uses an LED as a light source. Only one study investigated its use in the diagnosis of intestinal schistosomiasis, and it performed reasonably well, with a sensitivity of 85% and specificity of 96%.

7.4 Cyscope

Several studies focused on the use of Cyscope for the diagnosis of *Plasmodium* species, all of which were conducted in Africa. This device is a portable battery-operated fluorescent microscope with long life even in high power modes. Practically classified as a rapid detection test (RDT) for the diagnosis of malaria due to its ability to provide a result in under 10 min, it has the advantage of also being able to quantify parasitemia from blood specimens. Although the technology has been used in the diagnosis of tuberculosis, we did not include those articles here (Chaidir et al., 2013; Chang et al., 2015). Additionally, studies performed in dense population centers with greater resources were also excluded (Ogouyèmi-Hounto et al., 2013). We identified five studies investigating this technology in rural and low-resource settings in Africa. Except for one study (Sousa-Figueiredo et al., 2010), the device showed high sensitivity (>90%) and specificity (>87%) in the diagnosis of *Plasmodium* species, mostly *P. falciparum* (Hassan et al., 2010, 2011; Nkrumah et al., 2011; Birhanie, 2016). The exceptional study (Sousa-Figueiredo et al., 2010) had very high rates of false positives for unclear reasons.

8. DISCUSSION

Our review is the first to systematically search the available literature to determine the effectiveness of handheld and mobile phone devices for the microscopic diagnosis of malaria and NTDs in low-resource settings. While this field of research is rapidly evolving and the technologies are expected to be modified and improved in the near-future, our review provides a baseline for future systematic assessment of the topic. Our article builds on one prior review focused mobile phone devices in the diagnosis of parasitic diseases that included studies in controlled laboratory settings (Saeed and Jabbar, 2018).

Overall, the quality of the articles selected in our review were low in bias on the basis of the QUADAS2 criteria, a widely used tool for the assessment of diagnostic tests (Whiting, 2011). For example, all articles chose gold standard comparisons that are well-accepted in the field of microscopy, including thick Giemsa stains for the diagnosis of malaria and the Kato-Katz thick smear for the diagnosis of *S. mansoni* and STHs. Additionally, with regard to patient selection, studies generally chose random subjects from general populations in endemic settings to study rather than individuals with suspected high or low burdens of disease. Such a wide distribution of individuals lends to the validity of using these devices in the field. Thirdly, for nearly all samples studied in these investigations, the authors reported that both the experimental tool and index standard were tested and few samples were lost or remained untested. A notable exception to the high quality features of the articles were that three of the five studies examining the Cyscope did not blind their microscopists to the results of the reference standard. This lack of blinding may threaten the validity of some of these findings.

We found that the wide range of test performance characteristics varied depending on organism and setting. Notable examples of effective devices included the high specificity of CellScope in the diagnosis of *Schistosoma* species and *Loa loa*; the high sensitivity and specificity of the Newton Nm1 microscope for the diagnosis of schistosomiasis and STH; and the generally high specificities of using fluorescent microscopy (Cyscope) for the diagnosis of malaria. Although positive and negative predictive values are summarized in our review, it is important to interpret these values based on local disease prevalence. Thus, high predictive values in one setting may not translate to other settings with differing disease prevalence.

There are several important implications from our findings. Firstly, reasonable evidence exists that the Foldscope provides excellent specificity but low sensitivity in the diagnosis of *S. haematobium*. Secondly, the Loa-Scope has shown promise in the diagnosis of *Loa loa* infection in highly endemic regions of Africa. Because rapid and accurate diagnosis of *Loa loa* is also critical to the treatment and elimination of onchocerciasis in areas where these two parasites are co-endemic, this device may have a dramatic impact on public health programs in various regions across Africa. Thirdly, based on numerous studies, the Cyscope appears to be an accurate method of identifying malaria infection with rapid turnaround time in various low-resource settings, including hospitals and field clinics. This holds the advantage over traditional RDTs due to its ability to quantify parasitemia as well as reliably diagnose *P. falciparum* infections.

An important limitation to our review was the inability to conduct a mathematical analysis of the data to generate a summary measure of the effectiveness of the studied technologies. This is because the settings, diseases, diagnostic tools, and reference tests varied widely between the studies. Such an analysis may be possible once future additional research is performed to validate the use of these diagnostic tools for different diseases in several settings.

While the current study only evaluated devices tested in real-world low-resource settings, there are many new diagnostic tools that have demonstrated potential utility in early, laboratory-based proof-of-concept studies. Several creative approaches have recently been taken to image pathogens, including tomography, fluorescence, and holographic imaging modalities (Seo et al., 2009; Isikman et al., 2011; Zhu and Ozcan, 2013; Tapley et al., 2013; Zuo et al., 2015). Additionally, the use of high resolution wide-field lenses may be a more effective mechanisms to scan samples (Zhu et al., 2011). One device harnesses the buoyant properties helminth eggs to concentrate them in a small area and visualize them in a single field of view (Sowerby et al., 2016), and another device detects the movement of miracidia for schistosomiasis diagnoses (Linder et al., 2016). While these creative approaches demonstrate promise in the field of mobile microscopy, the next major step will be to transform them into modalities that can be evaluated in field settings with comparison to gold standard diagnostic tests.

Many of the devices discussed here rely on innovative imaging technology for users to better visualize samples. However, the field of handheld and mobile phone microscopy is now angling towards a promising direction where software may help visualize, identify, and quantify specific organisms independent of human observers. For example, machine learning may permit better image contrast to help identify organisms on mobile phones in addition to automating the diagnosis of giardiasis (Ceylan Koydemir et al., 2017), eggs of STH and *Schistosoma* (Linder et al., 2013; Slusarewicz et al., 2016; Holmström et al., 2017), and malaria parasites (Rosado et al., 2017). Much of the software has already been integrated onto the mobile phones for diagnosis, which may be a challenge for remote settings with limited local hard-drive storage and computing power. However, harnessing the power of cloud-based analyses and advancing the concept of telemedicine may be a solution to this barrier, whereby data processing may occur distant from where images are captured, and results are rapidly and electronically relayed back to front-line personnel (Linder et al., 2008).

An important aspect of successful mobile phone and handheld microscope technologies must be its ability to accommodate a broad array of samples and be sturdy enough to maintain high-level diagnostic performance in rugged conditions. While many of these devices are being designed for use in low-resource settings, more attention on the preparation of samples in such settings is required such that rapid point-of-contact diagnoses can be made (Rajchgot et al., 2017; Bogoch et al., 2017b). An interdisciplinary approach to device design and evaluation will likely result in better microscopes that are useful in real-world settings.

9. CONCLUSION

Handheld microscopes and mobile phone devices have the potential to improve the quality of clinical care and public health care for NTDs in low-resource settings. Current devices are slowly being integrated into routine practice and newer innovations such as computer vision and machine learning may aid in automating diagnoses. The latter holds great promise in bringing equitable laboratory diagnostic support to areas most in need. The field continues to rapidly evolve as more novel lens and lighting technologies are being developed and studied. Future investigation in the area will likely result in more accurate, portable, and economical tools used in the diagnosis of NTDs.

SUPPLEMENTARY DATA

Supplementary data related to this article can be found online at https://doi.org/10.1016/bs.apar.2018.09.001.

Financial Support
AV is funded by the Elliot Philipson Scholarship Program provided by the Department of Medicine at the University of Toronto. JRS is Director of COUNTDOWN, an implementation research consortium funded by DFID_UK. IIB is supported by a generous donation from the Tesari Foundation.

Dedication
We would like to dedicate our article to the memory of a dear friend and recently departed colleague Rick Dickinson who has left us much enriched by his wisdom and wonderful legacy of innovative portable microscope designs.

REFERENCES

Bastawrous, A., Armstrong, M.J., 2013. Mobile health use in low- and high-income countries: an overview of the peer-reviewed literature. J. R. Soc. Med. 106, 130–142. https://doi.org/10.1177/0141076812472620.

Birhanie, M., 2016. Comparison of partec rapid malaria test with conventional light microscopy for diagnosis of malaria in Northwest Ethiopia. J. Parasitol. Res. https://doi.org/10.1155/2016/3479457.

Bogoch, I.I., Andrews, J.R., Speich, B., Utzinger, J., Ame, S.M., Ali, S.M., Keiser, J., 2013. Mobile phone microscopy for the diagnosis of soil-transmitted helminth infections: a proof-of-concept study. Am. J. Trop. Med. Hyg. 88, 626–629. https://doi.org/10.4269/ajtmh.12-0742.

Bogoch, I.I., Coulibaly, J.T., Andrews, J.R., Speich, B., Keiser, J., Stothard, J.R., N'goran, E.K., Utzinger, J., 2014a. Evaluation of portable microscopic devices for the diagnosis of *Schistosoma* and soil-transmitted helminth infection. Parasitology 141, 1811–1818. https://doi.org/10.1017/S0031182014000432.

Bogoch, I.I., Andrews, J.R., Speich, B., Ame, S.M., Ali, S.M., Stothard, J.R., Utzinger, J., Keiser, J., 2014b. Quantitative evaluation of a handheld light microscope for field diagnosis of soil-transmitted helminth infection. Am. J. Trop. Med. Hyg. 91, 1138–1141. https://doi.org/10.4269/ajtmh.14-0253.

Bogoch, I.I., Sayasone, S., Vonghachack, Y., Meister, I., Utzinger, J., Odermatt, P., Andrews, J.R., Keiser, J., 2016. Diagnosis of *Opisthorchis viverrini* infection with handheld microscopy in Lao people's democratic republic. Am. J. Trop. Med. Hyg. 94, 158–160. https://doi.org/10.4269/ajtmh.15-0525.

Bogoch, I.I., Koydemir, H.C., Tseng, D., Ephraim, R.K.D., Duah, E., Tee, J., Andrews, J.R., Ozcan, A., 2017a. Evaluation of a mobile phone-based microscope for screening of *Schistosoma haematobium* infection in rural Ghana. Am. J. Trop. Med. Hyg. 96, 1468–1471. https://doi.org/10.4269/ajtmh.16-0912.

Bogoch, I.I., Lundin, J., Lo, N.C., Andrews, J.R., 2017b. Mobile phone and handheld microscopes for public health applications. The Lancet. Public Health 2, e355. https://doi.org/10.1016/S2468-2667(17)30120-2.

Ceylan Koydemir, H., Feng, S., Liang, K., Nadkarni, R., Benien, P., Ozcan, A., 2017. Comparison of supervised machine learning algorithms for waterborne pathogen detection using mobile phone fluorescence microscopy. Nanophotonics. https://doi.org/10.1515/nanoph-2017-0001.

Chaidir, L., Parwati, I., Annisa, J., Muhsinin, S., Meilana, I., Alisjahbana, B., van Crevel, R., 2013. Implementation of LED fluorescence microscopy for diagnosis of pulmonary and hiv-associated tuberculosis in a hospital setting in indonesia. PloS One 8. https://doi.org/10.1371/journal.pone.0061727 e61727.

Chang, E.W., Page, A.-L., Bonnet, M., 2015. Light-emitting diode fluorescence microscopy for tuberculosis diagnosis: a meta-analysis. Eur. Respir. J. https://doi.org/10.1183/13993003.00978-2015.

Collier, J.A.B., Longmore, J.M., 1983. The reliability of the microscopic diagnosis of malaria in the field and in the laboratory. Ann. Trop. Med. Parasitol. 77, 113–117. https://doi.org/10.1080/00034983.1983.11811683.

Coulibaly, J.T., Ouattara, M., D'Ambrosio, M.V., Fletcher, D.A., Keiser, J., Utzinger, J., N'Goran, E.K., Andrews, J.R., Bogoch, I.I., 2016a. Accuracy of mobile phone and handheld light microscopy for the diagnosis of schistosomiasis and intestinal Protozoa infections in Côte d'Ivoire. PLoS Neglected Trop. Dis. 10 https://doi.org/10.1371/journal.pntd.0004768 e0004768.

Coulibaly, J.T., Ouattara, M., Keiser, J., Bonfoh, B., N'Goran, E.K., Andrews, J.R., Bogoch, I.I., 2016b. Evaluation of malaria diagnoses using a handheld light microscope in a community-based setting in rural Côte d'Ivoire. Am. J. Trop. Med. Hyg. 95 (95), 831–834. https://doi.org/10.4269/ajtmh.16-0328, 10.4269/ajtmh.16-0328.

D'Ambrosio, M.V., Bakalar, M., Bennuru, S., Reber, C., Skandarajah, A., Nilsson, L., Switz, N., Kamgno, J., Pion, S., Boussinesq, M., Nutman, T.B., Fletcher, D.A., 2015. Point-of-care quantification of blood-borne filarial parasites with a mobile phone microscope. Sci. Transl. Med. 7, 286re4. https://doi.org/10.1126/scitranslmed.aaa3480.

Dunning, K., Stothard, J.R., 2007. From the McArthur to the millennium health microscope (MHM): future developments in microscope miniaturization for international health. Microscopy Today 15, 18–21. https://doi.org/10.1017/S1551929500050963.

Ephraim, R.K.D., Duah, E., Cybulski, J.S., Prakash, M., D'Ambrosio, M.V., Fletcher, D.A., Keiser, J., Andrews, J.R., Bogoch, I.I., 2015. Diagnosis of *Schistosoma haematobium* infection with a mobile phone-mounted foldscope and a reversed-lens CellScope in Ghana. Am. J. Trop. Med. Hyg. 92, 1253–1256. https://doi.org/10.4269/ajtmh.14-0741.

Hassan, S.E.-D.H., Okoued, S.I., Mudathir, M.A., Malik, E.M., 2010. Testing the sensitivity and specificity of the fluorescence microscope (Cyscope®) for malaria diagnosis. Malar. J. 9, 88. https://doi.org/10.1186/1475-2875-9-88.

Hassan, S.E.-D.H., Haggaz, A.E.D., Mohammed-Elhassan, E.B., Malik, E.M., Adam, I., 2011. Fluorescence microscope (Cyscope®) for malaria diagnosis in pregnant women in Medani Hospital, Sudan. Diagn. Pathol. 6, 88. https://doi.org/10.1186/1746-1596-6-88.

Holmström, O., Linder, N., Ngasala, B., Mårtensson, A., Linder, E., Lundin, M., Moilanen, H., Suutala, A., Diwan, V., Lundin, J., 2017. Point-of-care mobile digital microscopy and deep learning for the detection of soil-transmitted helminths and *Schistosoma haematobium*. Glob. Health Action 10, 1337325. https://doi.org/10.1080/16549716.2017.1337325.

Isikman, S.O., Bishara, W., Sikora, U., Yaglidere, O., Yeah, J., Ozcan, A., 2011. Field-portable lensfree tomographic microscope. Lab Chip 11, 2222–2230. https://doi.org/10.1039/c1lc20127a.

Kamgno, J., Pion, S.D., Chesnais, C.B., Bakalar, M.H., D'Ambrosio, M.V., Mackenzie, et al., 2017. A test-and-not-treat strategy for onchocerciasis in Loa-Loa–endemic areas. N. Engl. J. Med. 377, 2044–2052. https://doi.org/10.1056/NEJMoa1705026.

Katz, N., Chaves, A., Pellegrino, J., 1972. A simple device for quantitative stool thick-smear technique in schistosomiasis mansoni. Rev Inst Med Trop De Sao Paulo 14, 397–400.

Kreindler, R.J., 2013a. Folded-optics Microscopes 1 of 2: The Nm1 (Newton Microscopes): Their Heritage.

Kreindler, R.J., 2013b. The Nm1 (Newton Microscopes): Part 2 of 2 an In-depth Examination and Comparison to Other Folded-optics Designs.

Linder, E., Lundin, M., Thors, C., Lebbad, M., Winiecka-Krusnell, J., Helin, H., Leiva, B., Isola, J., Lundin, J., 2008. Web-based virtual microscopy for parasitology: a novel tool for education and quality assurance. PLoS Neglected Trop. Dis. 2, e315. https://doi.org/10.1371/journal.pntd.0000315.

Linder, E., Grote, A., Varjo, S., Linder, N., Lebbad, M., Lundin, M., Diwan, V., Hannuksela, J., Lundin, J., 2013. On-chip imaging of *Schistosoma haematobium* eggs in urine for diagnosis by computer vision. PLoS Neglected Trop. Dis. 7 https://doi.org/10.1371/journal.pntd.0002547 e2547.

Linder, E., Varjo, S., Thors, C., 2016. Mobile diagnostics based on motion? A close look at motility patterns in the schistosome life cycle. Diagnostics 6. https://doi.org/10.3390/diagnostics6020024.

Longmore, J.M., 1983. Bedside haematology: new facility in general practice. Br Med J 287, 1849–1851. https://doi.org/10.1136/bmj.287.6408.1849.

Longmore, J.M., 1986. An atlas of bedside microscopy. J. Roy. Coll. Gen. Pract. Occasional paper 1–31.

Meade Instruments, Telescopes, Solar Telescopes, Binoculars, Spotting Scopes, www.meade.com [Date accessed: October 3, 2018].

Moher, D., Liberati, A., Tetzlaff, J., Altman, D.G., Group, T.P., 2009. Preferred reporting items for systematic reviews and meta-analyses: the PRISMA statement. PLoS Med. 6 https://doi.org/10.1371/journal.pmed.1000097 e1000097.

Murray, C.J.L., Vos, T., Lozano, R., Naghavi, M., Flaxman, A.D., Michaud, C., Ezzati, M., Shibuya, K., Salomon, J.A., Abdalla, S., Aboyans, V., Abraham, J., Ackerman, I., Aggarwal, R., Ahn, S.Y., Ali, M.K., Alvarado, M., Anderson, H.R., Anderson, L.M., Andrews, K.G., Atkinson, C., Baddour, L.M., Bahalim, A.N., Barker-Collo, S., Barrero, L.H., Bartels, D.H., Basáñez, M.-G., Baxter, A., Bell, M.L., Benjamin, E.J., Bennett, D., Bernabé, E., Bhalla, K., Bhandari, B., Bikbov, B., Bin Abdulhak, A., Birbeck, G., Black, J.A., Blencowe, H., Blore, J.D., Blyth, F., Bolliger, I., Bonaventure, A., Boufous, S., Bourne, R., Boussinesq, M., Braithwaite, T., Brayne, C., Bridgett, L., Brooker, S., Brooks, P., Brugha, T.S., Bryan-Hancock, C., Bucello, C., Buchbinder, R., Buckle, G., Budke, C.M., Burch, M., Burney, P., Burstein, R., Calabria, B., Campbell, B., Canter, C.E., Carabin, H., Carapetis, J., Carmona, L., Cella, C., Charlson, F., Chen, H., Cheng, A.T.-A., Chou, D., Chugh, S.S., Coffeng, L.E., Colan, S.D., Colquhoun, S., Colson, K.E., Condon, J., Connor, M.D., Cooper, L.T., Corriere, M., Cortinovis, M., de Vaccaro, K.C., Couser, W., Cowie, B.C., Criqui, M.H., Cross, M., Dabhadkar, K.C., Dahiya, M., Dahodwala, N., Damsere-Derry, J., Danaei, G., Davis, A., De Leo, D., Degenhardt, L., Dellavalle, R., Delossantos, A., Denenberg, J., Derrett, S., et al., 2012. Disability-adjusted life years (DALYs) for 291 diseases and injuries in 21 regions, 1990-2010: a systematic analysis for the global burden of disease study 2010. Lancet 380, 2197–2223. https://doi.org/10.1016/S0140-6736(12)61689-4.

Nkrumah, B., Acquah, S.E., Ibrahim, L., May, J., Brattig, N., Tannich, E., Nguah, S.B., Adu-Sarkodie, Y., Huenger, F., 2011. Comparative evaluation of two rapid field tests for malaria diagnosis: Partec rapid malaria Test® and Binax Now® malaria rapid diagnostic test. BMC Infect. Dis. 11, 143. https://doi.org/10.1186/1471-2334-11-143.

Ogouyèmi-Hounto, A., Kinde-Gazard, D., Keke, C., Gonçalves, E., Alapini, N., Adjovi, F., Adisso, L., Bossou, C., Denon, Y.V., Massougbodji, A., 2013. [Assessment of a rapid diagnostic test and portable fluorescent microscopy for malaria diagnosis in Cotonou (Bénin)]. Bull. Soc. Pathol. Exot. 106, 27–31. https://doi.org/10.1007/s13149-012-0264-7.

Rajchgot, J., Coulibaly, J.T., Keiser, J., Utzinger, J., Lo, N.C., Mondry, M.K., Andrews, J.R., Bogoch, I.I., 2017. Mobile-phone and handheld microscopy for neglected tropical diseases. PLoS Neglected Trop. Dis. 11 https://doi.org/10.1371/journal.pntd.0005550 e0005550.

Rosado, L., da Costa, J.M.C., Elias, D., Cardoso, J.S., 2017. Mobile-based analysis of malaria-infected thin blood smears: automated species and life cycle stage determination. Sensors 17. https://doi.org/10.3390/s17102167.

Saeed, M.A., Jabbar, A., 2018. "Smart diagnosis" of parasitic diseases by use of smartphones. J. Clin. Microbiol. 56 e01469–17. https://doi.org/10.1128/JCM.01469-17.

Seo, S., Su, T.-W., Tseng, D.K., Erlinger, A., Ozcan, A., 2009. Lensfree holographic imaging for on-chip cytometry and diagnostics. Lab Chip 9, 777–787. https://doi.org/10.1039/b813943a.

Slusarewicz, P., Pagano, S., Mills, C., Popa, G., Chow, K.M., Mendenhall, M., Rodgers, D.W., Nielsen, M.K., 2016. Automated parasite faecal egg counting using fluorescence labelling, smartphone image capture and computational image analysis. Int. J. Parasitol. 46, 485–493. https://doi.org/10.1016/j.ijpara.2016.02.004.

Snow, R.W., Sartorius, B., Kyalo, D., Maina, J., Amratia, P., Mundia, C.W., Bejon, P., Noor, A.M., 2017. The prevalence of *Plasmodium falciparum* in sub-Saharan Africa since 1900. Nature 550, 515–518. https://doi.org/10.1038/nature24059.

Sousa-Figueiredo, J.C., Oguttu, D., Adriko, M., Besigye, F., Nankasi, A., Arinaitwe, M., Namukuta, A., Betson, M., Kabatereine, N.B., Stothard, J.R., 2010. Investigating portable fluorescent microscopy (CyScope®) as an alternative rapid diagnostic test for malaria in children and women of child-bearing age. Malar. J. 9, 245. https://doi.org/10.1186/1475-2875-9-245.

Sowerby, S.J., Crump, J.A., Johnstone, M.C., Krause, K.L., Hill, P.C., 2016. Smartphone microscopy of parasite eggs accumulated into a single field of view. Am. J. Trop. Med. Hyg. 94, 227–230. https://doi.org/10.4269/ajtmh.15-0427.

Stothard, J.R., Rollinson, D., 2018. Chapter one - an important milestone in parasitology: Celebrating a hundred volumes of advances in parasitology. In: Rollinson, D., Stothard, J.R. (Eds.), Advances in Parasitology. Academic Press, pp. 1–27. https://doi.org/10.1016/bs.apar.2018.03.004.

Stothard, J.R., Kabatereine, N.B., Tukahebwa, E.M., Kazibwe, F., Mathieson, W., Webster, J.P., Fenwick, A., 2005. Field evaluation of the Meade Readiview handheld microscope for diagnosis of intestinal schistosomiasis in Ugandan school children. Am. J. Trop. Med. Hyg. 73, 949–955. https://doi.org/10.4269/ajtmh.2005.73.949.

Stothard, J.R., Nabatte, B., Sousa-Figueiredo, J.C., Kabatereine, N.B., 2014. Towards malaria microscopy at the point-of-contact: an assessment of the diagnostic performance of the Newton Nm1 microscope in Uganda. Parasitology 141, 1819–1825. https://doi.org/10.1017/S0031182014000833.

Tapley, A., Switz, N., Reber, C., Davis, J.L., Miller, C., Matovu, J.B., Worodria, W., Huang, L., Fletcher, D.A., Cattamanchi, A., 2013. Mobile digital fluorescence microscopy for diagnosis of tuberculosis. J. Clin. Microbiol. 51, 1774–1778. https://doi.org/10.1128/JCM.03432-12.

Vos, T., Abajobir, A.A., Abate, K.H., Abbafati, C., Abbas, K.M., Abd-Allah, F., Abdulkader, R.S., Abdulle, A.M., Abebo, T.A., Abera, S.F., Aboyans, V., Abu-Raddad, L.J., Ackerman, I.N., Adamu, A.A., Adetokunboh, O., Afarideh, M., Afshin, A., Agarwal, S.K., Aggarwal, R., Agrawal, A., Agrawal, S., Ahmadieh, H., Ahmed, M.B., Aichour, M.T.E., Aichour, A.N., Aichour, I., Aiyar, S., Akinyemi, R.O., Akseer, N., Lami, F.H.A., Alahdab, F., Al-Aly, Z., Alam, K., Alam, N., Alam, T., Alasfoor, D., Alene, K.A., Ali, R., Alizadeh-Navaei, R., Alkerwi, A., Alla, F., Allebeck, P., Allen, C., Al-Maskari, F., Al-Raddadi, R., Alsharif, U., Alsowaidi, S., Altirkawi, K.A., Amare, A.T., Amini, E., Ammar, W., Amoako, Y.A., Andersen, H.H., Antonio, C.A.T., Anwari, P., Ärnlöv, J., Artaman, A., Aryal, K.K., Asayesh, H., Asgedom, S.W., Assadi, R., Atey, T.M., Atnafu, N.T., Atre, S.R., Avila-Burgos, L., Avokphako, E.F.G.A., Awasthi, A., Bacha, U., Badawi, A., Balakrishnan, K., Banerjee, A., Bannick, M.S., Barac, A., Barber, R.M., Barker-Collo, S.L., Bärnighausen, T., Barquera, S., Barregard, L., Barrero, L.H., Basu, S., Battista, B., Battle, K.E., Baune, B.T., Bazargan-Hejazi, S., Beardsley, J., Bedi, N., Beghi, E., Béjot, Y., Bekele, B.B., Bell, M.L., Bennett, D.A.,

Bensenor, I.M., Benson, J., Berhane, A., Berhe, D.F., Bernabé, E., Betsu, B.D., Beuran, M., et al., 2017. Global, regional, and national incidence, prevalence, and years lived with disability for 328 diseases and injuries for 195 countries, 1990–2016: a systematic analysis for the Global Burden of Disease Study 2016. Lancet 390, 1211–1259. https://doi.org/10.1016/S0140-6736(17)32154-2.

Whiting, P.F., 2011. QUADAS-2: a revised tool for the quality assessment of diagnostic accuracy studies. Ann. Intern. Med. 155, 529. https://doi.org/10.7326/0003-4819-155-8-201110180-00009.

Zhu, H., Ozcan, A., 2013. Wide-field fluorescent microscopy and fluorescent imaging flow cytometry on a cell-phone. JVE. https://doi.org/10.3791/50451.

Zhu, H., Yaglidere, O., Su, T.-W., Tseng, D., Ozcan, A., 2011. Cost-effective and compact wide-field fluorescent imaging on a cell-phone. Lab Chip 11, 315–322. https://doi.org/10.1039/c0lc00358a.

Zuo, C., Sun, J., Zhang, J., Hu, Y., Chen, Q., 2015. Lensless phase microscopy and diffraction tomography with multi-angle and multi-wavelength illuminations using a LED matrix. Optic Express 23, 14314–14328.

FSC
www.fsc.org
MIX
Papier | Fördert
gute Waldnutzung
FSC® C083411

Kolibri 360 GmbH
Ferdinand-Jühlke-Straße 7
99095 Erfurt, Deutschland
produktsicherheit@kolibri360.de